OUTDOOR
EXPEDITION
BOOK 99

I AM BEAMS Vol.9 **SHIGERU KANEKO**

EXPEDITION —
冒険、探検、遠征の意味のあるこの言葉に、僕はアメリカを感じます。

この言葉に取り憑かれたのは、一着のダウンジャケットがきっかけでした。それはシエラデザインズが1968年にリリースしたミナレットパーカ。パンパンに詰め込まれたダウンとそのボリューミーなルックス、極寒地に挑むためのEXPEDITIONスペックが生む機能美を前にし、僕の洋服の概念は音を立てて崩れ、クラシックアウトドアの世界に完全に惚れ込んでしまったのです。それらは僕の洋服への考え方に大きな影響を与え、アイデアの源泉でもあります。あらゆる素材が、あらゆるディテールが、"冒険を成し遂げるため"にあるEXPEDITIONなギアたち。本書は、そのルーツと魅力を僕の秘蔵コレクションとともにEXPEDITIONしていく一冊です。

EXPEDITION is the word from which I feel America

My obsession with this word began with a down jacket. It is the Minaret Parka which Sierra Designs produced in 1968. From the down packed to the brim to the voluminous appearance, the functional beauty created by EXPEDITION technologies for extreme cold regions overturned my concept of clothing, and I fell completely in love with the world of classic outdoor gear. They wielded a big influence on my way of thinking about clothing and also have been inspiring me. Every single material and detail of EXPEDITION gear is designed to "accomplish an adventure". This book invites you to an EXPEDITION to explore the roots and the attraction by referring to my treasures.

金子 茂
Shigeru Kaneko
BEAMS PLUS / Chief Buyer

1984年生まれ。2010年に「ビームス 原宿」のアルバイトを経て入社。2015年に〈ビームス プラス〉のバイヤーに就任。古着への造詣が深く、ヴィンテージを元に再構築するものづくりやバイイングを得意とする。そのアウトドアコレクションは業界でも有名。

Born in 1984. He joined BEAMS in 2010 after working at BEAMS HARAJUKU as part-time staff. In 2015, he became a buyer for BEAMS PLUS. With the extensive knowledge of vintage clothes, he excels at making vintage-inspired clothes and buying. His collection of outdoor garments is well known within the fashion industry.

CONTENTS

P.6 **01** OUTDOOR RECREATION ARCHIVE

EXPEDITION PARKAS OF U.S.A.

P.18 **02** Eddie Bauer / MOUNT EVEREST
P.19 **03** Eddie Bauer / MOUNT EVEREST
P.20 **04** Eddie Bauer / DOWN PARKA
P.21 **05** Eddie Bauer / DOWN FLIGHT PANTS
P.22 **06** THE NORTH FACE / NORTH FACE PARKA
P.23 **07** THE NORTH FACE / BROOKS RANGE
P.23 **08** THE NORTH FACE / BROOKS RANGE
P.24 **09** Holubar / EXPEDITION PARKA
P.25 **10** Holubar / COLORADO PARKA
P.26 **11** Gerry / U.S.A.F SURVIVAL SUIT
P.27 **12** Gerry / WALKAROUND SLEEPER SHORT COAT
P.28 **13** Gerry / ANDEAN JACKET
P.29 **14** Gerry / ARCTIC JACKET
P.30 **15** Snow Lion / EXPEDITION PARKA
P.31 **16** Snow Lion / BERING PARKA
P.32 **17** SIERRA DESIGNS × BEAMS PLUS / MINARET PARKA
P.33 **18** SIERRA DESIGNS × KAPTAIN SUNSHINE × BEAMS PLUS / MINARET PARKA
P.34 **19** Recreational Equipment, Inc. / SUMMIT PARKA
P.35 **20** Recreational Equipment, Inc. / JAMET DOWN PARKA
P.36 **21** Recreational Equipment, Inc. / DENALI EXPEDITION PARKA
P.37 **22** Recreational Equipment, Inc. / EXPEDITION BIVOUAC PANTS
P.38 **23** Alpine Designs / EXPEDITION PARKA
P.39 **24** Frostline Kits / TUNDRA JACKET
P.40 **25** BEAMS PLUS / EXPEDITION PARKAS
P.42 **26** Adventure 16 / WIDE TEMPERATURE RANGE PARKA
P.43 **27** WILDERNESS EXPERIENCE / ULTIMATE PARKA

P.44 **28** EXPEDITION DOWN PARKAS SPECIAL TALK with SHINSUKE NAKADA

EXPEDITION PARKAS OF EUROPE

P.52 **29** Blacks / DUVET JACKET
P.54 **30** MONCLER / DUVET JACKET
P.55 **31** MONCLER / KARAKORUM
P.56 **32** Pointfive / DUVET JACKET
P.57 **33** Pointfive / DUVET JACKET
P.58 **34** MOUNTAIN EQUIPMENT/CERRO TORRE
P.59 **35** Himalchuli / DUVET JACKET

FIELD JACKETS

P.62 **36** L.L.Bean / BEAN'S DOWN MACKINAW
P.63 **37** WOOLRICH / 8PT BARRIER HUNTING COAT
P.64 **38** Alyeska / PIPELINE DOWN COAT
P.65 **39** Patagonia / DOWN SWEATER
P.66 **40** WOOLRICH / HUNT PARKA
P.67 **41** Marmot / THE PICKET PIN PARKA
P.68 **42** Gerry / SLOPE COAT
P.69 **43** White Stag / SKI JACKET
P.70 **44** Early Winters / THE LOST WORLD PARKA
P.71 **45** Levi's / DENIM MOUNTAIN PARKA

EDDIE BAUER ITEMS

P.74 **46** Eddie Bauer / DOWN ACCESSORIES
P.75 **47** Eddie Bauer / DOWN WEARS

P.76 **48** DOWN VESTS SPECIAL TALK with MANABU HARADA

CAGOULES

P.84 **49** THE NORTH FACE / CAGOULE
P.85 **50** Sierra West / STORM KING CAGOULE
P.86 **51** Alpine Designs / NYLON CAGOULE
P.86 **52** Gerry / GERRY PARKA
P.87 **53** Blacks / ANORAK VENTILE MODEL
P.87 **54** Recreational Equipment, Inc. / CO-OP ANORAK

FIELD PARKAS

P.90 **55** THE NORTH FACE made by SIERRA DESIGNS / STANDARD MOUNTAIN PARKA
P.91 **56** SIERRA DESIGNS / MOUNTAIN PARKA REEVAIR MODEL
P.91 **57** KELTY / MACKINTOSH ALL WEATHER PARKA
P.92 **58** Gerry / MOUNTAIN PARKA
P.92 **59** Recreational Equipment, Inc. / MOUNTAINEER STORM PARKA
P.93 **60** Early Winters / TERRASHELL
P.93 **61** WOOLRICH / MOUNTAIN PARKA
P.94 **62** L.L.Bean / SUMMIT PARKA
P.94 **63** Alpine Hut / SKI PARKA
P.95 **64** Survivalon / SAILING JACKET
P.95 **65** Rocky Mountain Featherbed / MOUNTAIN JACKET

THE OTHER MY COLLECTIONS

P.98 **66** LONG BILL CAPS
P.100 **67** FIELD HATS
P.101 **68** Willis & Geiger / SAFARI
P.102 **69** HUNTER YELLOW · SAFETY YELLOW
P.104 **70** FILSON / MACKINAW CRUISER
P.105 **71** Hercules / DUCK COATS
P.106 **72** L.L.Bean / OLD L.L.Bean
P.107 **73** L.L.Bean / RAINBOW LAKE JACKET & PANTS
P.108 **74** Red Head / LINER VESTS
P.109 **75** POLISH ARMY CHEETAH CAMOUFLAGE
P.110 **76** LINER JACKETS
P.111 **77** Bradley Mountain Wear / FLEECE JACKETS
P.112 **78** Early Winters / GORE TRAINER SUITS
P.113 **79** L.L.Bean / REFLECTOR CYCLING JACKET
P.114 **80** SKI
P.115 **81** Mary Maxim / COWICHAN SWEATERS

P.116 **82** GEAR VESTS SPECIAL TALK with SETSUMASA KOBAYASH

P.122 **83** L.L.Bean × BEAMS PLUS / DEEP BOTTOM BOAT AND TOTE
P.123 **84** L.L.Bean × BEAMS PLUS / BEAN BOOTS
P.124 **85** GREGORY × KAPTAIN SUNSHINE × BEAMS PLUS / CASSIN
P.125 **86** SPORTS SHIRTS
P.126 **87** PATCHES
P.127 **88** Recreational Equipment, Inc. / PONCHO
P.128 **89** SIERRA DESIGNS × BEAMS PLUS / EXCLUSIVE PRODUCTS
P.129 **90** DOWN SHOES
P.129 **91** DOWN HOOD
P.130 **92** Eddie Bauer / DOWN FACE MASKS
P.131 **93** BALACLAVAS
P.132 **94** MOCCASIN SHOES
P.133 **95** CATALOGS
P.134 **96** Frostline Kits / KITS
P.136 **97** Banana Republic / TRAVELER'S SPORTS COAT
P.137 **98** Patagonia / BLAZER & BOMBACHAS

P.138 **99** LOG HOUSE

5

•TRIP to THE OUTDOOR RECREATION ARCHIVE

01 HELLO, UTAH

TRIP TO THE OUTDOOR RECREATION ARCHIVE

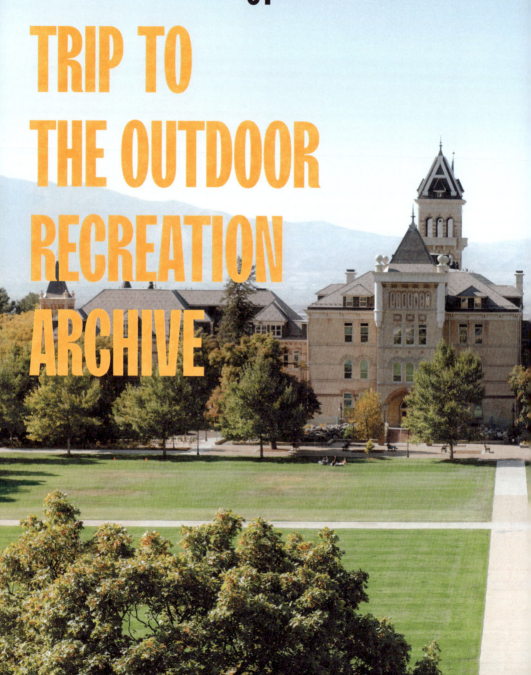

ユタ州立大学の図書館
Outdoor Recreation Archiveに行ってきた!

山がモチーフのロゴでお馴染みのグレゴリー本社をはじめ、米・ユタ州にはアウトドアギアのメーカーやショップが数多くひしめいている。そのアウトドアの聖地で活動する、Outdoor Recreation Archive (以下ORA) という組織をご存じだろうか。自然公園に囲まれた緑豊かな都市、ローガンにキャンパスを構えるユタ州立大学内で、アウトドア関連の膨大な資料を収蔵しているORA。今回、金子独自のネットワークにより、現地での取材が実現した。

We have visited the Outdoor Recreation Archive in Utah!

Many outdoor companies and retail shops locate their business operations in Utah, USA. In such an outdoor-oriented place, the Outdoor Recreation Archive (ORA) stores and gives access to a vast collection of historical materials of outdoor gear as a part of Utah State University (USU) in Logan, a city surrounded by natural parks. This time, our visit to the organization was enabled by Kaneko's personal network.

●TRIP to THE OUTDOOR RECREATION ARCHIVE

ORA 01 DOOR TO THE HUGE COLLECTION

Clint Pumphrey (Left)
クリンド・パンフリー
Library Archivist

大学図書館のスペシャル・コレクションズ&アーカイブスの司書であり、ORAの業務にも携わる。趣味はハイキングやトレイルラン。

Clint serves the ORA, working as an archivist in the library's Special Collections and Archives department. Enjoys hiking and trail running.

Chase Anderson (Right)
チェイス・アンダーソン
Industry Relations Manager

アウトドア業界での職務経験を経て、アウトドア・プロダクト・デザイン学部の職員に転身。ORAでは主にSNS発信などの宣伝広報を担当する。

After working in outdoor industry, Chase joined the Outdoor Product Design & Development program. At the ORA, he is in charge of social media manegement and public relations.

アウトドア関連の膨大な資料を所蔵する

2005年に建て替えられたユタ州立大学の図書館。ユタ州にまつわる詳細な資料や学部関連の書物を大量に収蔵しているが、その中でも貴重な資料を保管している特別な部屋がある。名称は"スペシャル・コレクションズ&アーカイブス"。学生はデータから必要な資料を探し、それを担当者に伝えると奥の資料室から取り出してきてくれる。学生はその部屋でのみ閲覧できるという仕組みだ。ORAの資料は、その奥深くの秘密の部屋で厳重に保管されている。

A vast collection of outdoor-related materials

USU's library was rebuilt in 2005. Along with a variety of materials about Utah and university-related books, the huge collection consists of a lot of valuable materials stored in a specially dedicated room named Special Collections and Archives. To access materials in the room, students need to search for them online before requesting library staff to retrieve them from the backroom. The materials can be read and used only within the Special Collections and Archives room. The ORA stores its materials in a secret room at the end with utmost care.

ORAを運営するキーマンはアウトドアカルチャーへの造詣が深い2人。アウトドア・プロダクト・デザイン学部に従事するチェイス氏と、図書館の司書であるクリント氏である。実はORAと金子はインスタグラムで繋がっており、それが今回のユタ取材の端緒となった。というのも、金子と彼らはすでに一つのプロジェクトを共にしている。『THE OUTDOOR ARCHIVE』という、ORAの資料をベースにした彼らのアートブックに、金子が寄稿しているのだ。SNSを通じて生まれた数奇な絆から、初対面を叶える機会を得たというわけだ。

There are two key persons involved in the ORA: Chase Anderson in the Outdoor Product Design & Development faculty and Clint Pumphrey, a librarian who is also well versed in outdoor culture. The ORA and Kaneko mutually follow each other on Instagram and that led us to embark on this trip to Utah. It was actually not the first time for them to collaborate. Kaneko has contributed an article to "THE OUTDOOR ARCHIVE", an art book Chase and Clint edited with materials from the ORA. And this time they could finally have the first opportunity to meet each other in person.

● TRIP to THE OUTDOOR RECREATION ARCHIVE

ORA 01 DIGGING DEEP INTO THE ARCHIVE

大量&貴重なアーカイブを 2日間ひたすらディグ

ORAが持つ資料は膨大だ。その数、雑誌・書籍でおよそ1万冊、ブランドカタログは6500冊に及ぶという。アメリカの雑誌はもちろんのこと、驚くことに『Made in U.S.A. catalog』『山の道具手帳』といった日本の名著も。カタログも、グレゴリー、パタゴニア、ザ・ノース・フェイス、エルエルビーンといった著名ブランドから、ビームスのシーズンカタログまで！ その収集力とネットワークたるや恐るべし、である。

Exploring the treasured archive for 2 days

The ORA owns an enormous number of materials. Their collection of 10,000 books and magazines consists of not just American publications but also Japanese ones such as "Made in U.S.A. catalog" and "Yama no Dogu Techo". As for catalogs, they have about 6,500 of those that include ones from renowned brand like Gregory, Patagonia, The North Face and L.L.Bean. And they even own BEAMS catalogs! Their network and ability of collecting are just outstanding.

今回のユタ訪問における最大の目的は、本書に掲載するプロダクトに関するリサーチと事実確認であった。ORAには事前に金子から見たいブランドのカタログをリクエストさせてもらい、到着時には万全の態勢で迎えてくれる厚遇ぶり(感謝！)。まる2日間、図書館にこもりきってひたすら資料を読みあさるという、この上ない至高の時間を得られた。とはいえ、眼前には貴重な資料が文字通り山ほど広がっている。最低限見るべきものは見られたものの、お宝の山を前に金子がこぼした一言は「時間が全然足りない(苦笑)」だった。

The main purpose of our Utah trip was to research the products featured on this book and verify the facts. We had sent the ORA a list of brands' catalogs that Kaneko requested in advance, and they kindly prepared everything on the sidelines of our visit. We do appreciate for that! Kaneko then spent whole two days in the library to devour those catalogs and other materials. It was truly a blissful, unforgettable experience for him to be surrounded by so many valuable catalogs. There are, however, too many materials to see and he could look through the least number of them. "I hardly have enough time…" Kaneko complained so but with a smile.

●TRIP to THE OUTDOOR RECREATION ARCHIVE

ORA 01 TALKING ABOUT THE "ORA"

学生、ユタ州、アウトドアカルチャー。それらに対する未来への投資

なぜ、ORAはこれほどのアウトドア資料を集められるのか。しかも非常に高いモチベーションで。その理由は、チェイス氏が籍を置くアウトドア・プロダクト・デザイン学部にある。ユタ州からの要請を受け、アパレルからギアまでアウトドア関連全ての企画、デザイン、製作、プロモーションに関わるA to Zを学べる、全米初の学部として発足したのが2015年。ORAの活動開始はその2年後で、その目的は大学＝育成機関としての役割にある。

Supporting students, Utah and outdoor culture for the future

Why can the ORA collect so many materials related to outdoor culture with such enthusiasm? USU's Outdoor Product Design & Development faculty to which Chase belongs has the reason. Since starting its history in 2015 at the request from Utah's state government, it has been the first-ever place in the U.S. to study the planning, design, manufacturing and PR of all outdoor products from clothing to gear. The ORA was launched two years later to help students study further.

Clint: "Our first priority is to provide a resource for our students. We want them to develop an appreciation for the history of the outdoor industry

「最優先は学生たちにリソースを提供すること。将来アウトドア業界に進むため、この産業に対する歴史認識を深めてもらう。最先端の技術と知識に加えてヴィンテージのスタイルやトレンドを理解した彼らは、ここユタ州や世界中で革新を起こせるはず。そういう意味で未来への投資なんです」（クリント氏）

「学生だけでなく業界関係者もORAの価値を認めています。クリエイターやマーケターから、インスタをフォローされたり資料の閲覧を求められたり。これら資料を保存し、発信することも使命の一つです」（チェイス氏）

that they can take with them into their outdoor-related careers. These well-rounded students, armed with an understanding of both the cutting edge skills involved in making modern clothing and gear as well as the influence of vintage styles and trends, have the tools to make the outdoor industry more innovative and prosperous here in Utah and across the globe. In this sense, we are using the past to invest in the future." Chase: "In addition to our students, industry has found the ORA incredibly valuable. Designers, creatives, marketers, and more frequently follow the Instagram account, contact Clint requesting access to materials, or visit us in person to conduct design research in the collection. Preserving and making these materials more publicly available is one of our missions!"

● TRIP to THE OUTDOOR RECREATION ARCHIVE

ORA **01** # THANK YOU VERY MUCH, UTAH!

ユタ州ローガンは
最高のアウトドア都市

ユタ州立大学内だけでなく、ローガンという
都市とそこに住む人々には、アウトドアカル
チャーが芯から根付いている。スキーやスノ
ーボードといったスノースポーツ、山や渓谷
でのハイクやトレイル、そのほかにもロード
バイク、フィッシング、はたまたハンティン
グなども日常の一部となっているのだ。100年
以上の歴史をもつ大型アウトドアギア店、店
主が個人でロードバイクレースを主宰するサ
イクリング店と、ショップも実にユニークだ。

Logan, Utah. One of
the best outdoor destinations

Not just at Utah State University, the city of Logan
and the people living there are deeply connected to
outdoor culture. All kinds of activities, from skiing,
snowboarding, hiking, trail running, road biking,
fishing and hunting, are a part of their everyday life.
There are also some unique retailers in the city,
including a large outdoor gear store with over 100
years of history and a cycling shop owned by a man
who personally hosts a road bike race.

取材で訪れたORAの本拠地であるユタ州立大学のキャンパスでも、学生が自転車やスケートボードで悠々と移動していたり、広々とした芝生でサッカーボールを蹴っていたりする姿が印象的だった。周囲を豊かな緑に囲まれたキャンパスで、桁違いに雄大な自然とおおらかな国民性の中で育まれたアメリカらしいアウトドアカルチャーを体感できたことは、金子自身のアウトドアファッションとの向き合い方や表現方法にもきっと好影響をもたらしてくれることだろう。改めて感謝を伝えたい。ありがとう、ORA！ ありがとう、ユタ！

At USU, we saw some students riding a bike and skateboard, while some others played soccer in a big grassy area. In a campus surrounded by greenery, Kaneko had a great first-hand experience to feel American outdoor culture fostered by the magnificent nature and generous-hearted national character. That will have a positive impact on him, his expression and his attitude towards fashion. We would like to express our gratitude again. Thank you very much, the ORA! Thank you very much, Utah!

EXPEDITION PARKAS OF U.S.A.

素材に、ディテールに、目を見張る 個性派揃いのアメリカン

高山がそこまで多くないアメリカでエクスペディションダウンジャケットが進化、発展した背景には、極寒の地であるアラスカやミリタリーウェアとしての需要が挙げられるでしょう。1960年代の半ばには、ウエストコーストを中心にダウンウェアを扱うさまざまなブランドが誕生します。アメリカのアパレルは生地やジップの開発に長けていたこともあり、最先端のスペックのエクスペディションダウンジャケットが次々に誕生しました。オリジナリティにあふれる製品が多いのも、アメリカのエクスペディションダウンの面白さです。僕が着ているダウンジャケットは、ザ・ノース・フェイスの名作ブルックスレンジ。中に着込んだクラシックなラガーシャツをポイントにしつつ、スウェットパンツやニューバランスを合わせて、アメリカンスタイルのリラックスムードを添えました。

Uniquely made jackets from the United States

Expedition parkas have been improved and popularized in the US, although the country doesn't have many high mountains. There are a few possible reasons for this; Alaska and military demands. A variety of brand manufacturing down-filled garments had been established by the mid 1960s mainly in the West Coast and invented the top-notch expedition down jackets because American companies exceled at developing quality fabrics and zippers. Uniqueness is one of the interesting characteristics of American-made expedition down jackets. The one I'm wearing is famous Brooks Range jacket from The North Face. I pair sweat pants and New Balance sneakers with a classic rugby shirt for a relaxing American feel.

● EXPEDITION PARKAS of U.S.A.

Eddie Bauer 02 MOUNT EVEREST

1963年、エベレスト登頂を目指す登山家に向けて開発されたダウンジャケットです。極寒のエベレスト用だけあってダウン量が群を抜いて多く、着るともうパンパン！ カタログを見ると、カラー名はスカーレット、ファーはウルヴァリン(イタチ科の動物)ファーとあります。それにしても、モデル名がマウント・エベレストって。顧客へのアピールが安直、もとい、直球なあたりも可愛げがあって好きだなぁ。

This down jacket was designed in 1963 for alpinists who aim for Mt. Everest. Made for mountaineering in extreme conditions, it is filled with enough down to make me look like a balloon! The catalog says the color is named "scarlet" and wolverine fur is attached. And the product itself is called "Mount Everest"? I think that's too straightforward, and quite simple but I like it.

Eddie Bauer 03 MOUNT EVEREST

パンパンのダウンに、ジップ＆ベルクロ留めのフロント。デザインは間違いなくマウント・エベレストなのですが、ファーが付いていません。テストサンプルなのか、個人オーダーなのか。カタログにも載っておらず、ナゾなだけに余計に思い入れのある一着です。セージグリーンの淡い色味も気に入っていて、時折無性に着たくなります。

Designed with a bulky shell filled with plenty of down and a zipped/Velcro closure, it looks just like the brand's Mount Everest jacket but comes without fur. Though I'm not sure whether it was made as a prototype or for a custom order, since it's not shown in catalogs. However the mysterious background makes me love this jacket more. The soft sage color is also my favorite and I sometimes become eager to wear it without any reason.

● EXPEDITION PARKAS of U.S.A.

Eddie Bauer 04 DOWN PARKA

第二次世界大戦の初期に、USエアフォースが採用していたダウンフライトスーツ。そもそもは30年代後半に極寒地アラスカの任務に当たるブッシュパイロットのために開発されたモデルですね。生地は今で言うワックスドコットン。エディー・バウアーの軍支給品としてはこれが最初のモデルで、ジャケットはこの後に登場するB-9へと移行する間のカタチです。デッドグラスカラーのボディと補強布のカラーコンビネーションが最高!!

Eddie Bauer 05 DOWN FLIGHT PANTS

This down flight suit was adopted by the U.S. Air Force during the early World War II period. It was actually first invented for bush pilots serving in extremely cold Alaska in the end of 1930s. The fabric is what we now call waxed cotton. The suit was the first product Eddie Bauer designed for military use, and the jacket is the predecessor model of the B-9. The color combination of the dead grass colored shell and the reinforcement fabric is just amazing!!

● EXPEDITION PARKAS of U.S.A.

THE NORTH FACE 06 NORTH FACE PARKA

心を揺さぶる、アカプルコゴールドというこの黄色。70年代初期〜半ばのモデルですが、後にこの黄色はなくなってしまうのでレアなんですよ。特大の丸いベルクロや当時の寝袋に用いられていたコイルジップに趣を感じます。モデル名に芸がないなと思いカタログを見たら、"アラスカパーカに改名したい"と書いてありました(笑)。

This stunning yellow color is called Acapulco Gold. This jacket had been produced from the early to mid 1970s, but the color was discontinued. So, this piece here is quite rare. The huge Velcro closures which coiled zipper often used for sleeping bags back then look tasteful. I was thinking the name of the jacket is too straightforward when looking through the catalog, but also found that it says the brand wants to change the name to Alaska Parka.

THE NORTH FACE 07　08 BROOKS RANGE

70年代後半に登場したモデルで、モデル名はアラスカのブルックス山脈から。カタログには"犬ぞりや氷点下の作業にも最適"とあるので、まさにアラスカのためのエクスペディション仕様ですね。左頁の一着とディテールはほぼ一緒ですが、こちらは表地がナイロンではなくてポリエステルコットンの65／35ベイヘッドクロス。アメリカらしい生地のアップデートがたまらない。

This model was first designed in the end of 1970s and the name derives from the Brooks Range in Alaska. As the catalog says it is ideal for dog sledding or work in a below-freezing environment, it's truly an expedition jacket. Though the detailing looks almost same as the one on the left page, the fabric of these ones are the 65/35 bayhead cloth. Such a fabric improvement is often seen on American-made products and I love it.

● EXPEDITION PARKAS of U.S.A.

Holubar 09 EXPEDITION PARKA

軍余剰品の寝袋の販売に端を発する、1947年スタートのコロラドのアウトドアブランドより。こちらは70年代初期のアルパインシリーズ最高峰モデルです。ホルバーが誇る、寝袋の最上級モデルROYALIGHTシリーズの性能をジャケットにしたものですが、ダウン量がとにかく凄くて、パンパンでしょう？ フロントはダブルスナップボタン&ジップ留め。冷気が入ってこないようオフセットが特に深くなっているのも特徴的です。

Holubar is a Colorado-based outdoor wear brand launched in 1947 to sell military surplus sleeping bags. This jacket is the highest-grade model of the Alpine series released in the early 1970s. It comes with the same features of the brand's premium sleeping bag series called Royal Light and is filled with a bulging amount of down. With a dual press stud/zipped closure, the front flaps are deeply overlapping to prevent cold air from entering.

Holubar 10 COLORADO PARKA

ブランド発祥の地を冠するコロラドパーカ。左頁のセカンドグレードモデルですが、そのスペックと実績は圧巻。ダブルチューブ製法により−30℃の寒さまで対応することができます。アメリカ自然保護団体シエラクラブで採用され、4ヶ月にも渡るネパール、ヒマラヤ、シベリアの6000マイルの旅に耐え抜いた実績も。ロマンを掻き立てられる一着です。

Colorado Parka is named after the birthplace of the brand. Even though this is the second grade model of the one on the left page, its specifications and track records are outstanding. Featuring double-tube system, it can withstand -30℃. It was proven when Sierra Club, an American environmental organization, endured a 6,000-mile journey for four months through Nepal, Himalaya to Siberia. This piece that sparks romance and adventure.

● EXPEDITION PARKAS of U.S.A.

Gerry 11 U.S.A.F SURVIVAL SUIT

米空軍から依頼を受けてヒマラヤ遠征隊に向けて開発された、寝袋を兼ねたダウンスーツです。マイナス50℃の環境下において、着用しながら活動できるというアンビリーバブルなアイテム。生産が開始された1959年には300着を工場フル稼働で縫いあげたと記録があります。大変貴重なので、大切にしたいですね。

Originally requested by the U.S. Air Force, this down suit, which also functions as a sleeping bag, was first developed for the Himalayan Expedition. Although designed to withstand a -50℃ environment, you can walk around in it. How unbelievable! In 1959, when production began, records show 300 suits were sewn at full factory capacity. Because this piece is valuable, I will take good care of it.

Gerry 12 WALKAROUND SLEEPER SHORT COAT

カタログの記述を見る限り、左頁のダウンスーツをセパレート仕様にアップデートして民間用に発売したモデルだと推測します。見頃から袖まで走ったダウンステッチが、寝袋からウェアになった瞬間を感じさせてくれる一着です。

As far as I can tell from reading the catalog, I assume this could be a model released for civilian use, updating the down suit on the left page to a separate design. The down stitching runs from the body to the sleeves, creating a garment that gives the feeling of a sleeping bag turning into wear.

27

● EXPEDITION PARKAS of U.S.A.

Gerry 13 ANDEAN JACKET

1964年に登場したジェリーの最高峰ダウンジャケット。キャッチフレーズは"丈夫で、暖かく、そして飛べるほど軽い"。ダウン量もさることながら、フロントに施された2列のジッパーが魅力的です。そこに上下から開閉可能なジッパーディテールも相まって、いわば"ダブルダブルジッパー"ですね。まさにジッパー発祥のアメリカらしいデザイン。冷気を防ぐその製法も流石の一言です。

Gerry launched this premium down jacket in 1964. With a catchphrase "Tough, warm & light enough to fly", it is filled with plenty of down and features appealing double zip front closure. The two-way zippers can be opened both upwards and downwards. So I may be able to call it "a double-double zip closure". The design is typical of America where zippers originate. Their manufacturing method for preventing cold air is truly impressive, too.

Gerry 14 ARCTIC JACKET

マスクのように顔を覆えるフードの作り が面白い、70年〜80年代に作られたモ デル。当時のカタログには、"最も過酷 な遠征条件に対応できるよう設計され た"とあります。そんなわけなので、ダ ウン量はパンパン。フロントはダブルス ナップ&ジップ留めで、まさかの袖にも ジップが付く仕様。ジェリーらしさあふ れる、ディテールに惚れる名品ですね。

Designed with a unique hood covering the face like a face mask, this jacket was produced in the 1970-1980s and described as "Quality designed and crafted for even the toughest expedition conditions," in the catalog back then. That is why it is plumped up with plenty of down. Along with the dual press stud/zipped front closure, the sleeves also feature zippers. Such a detailed design is often seen on Gerry's products. Love it.

●EXPEDITION PARKAS of U.S.A.

Snow Lion 15 EXPEDITION PARKA

バークレーで設立されたブランドの一着で、日焼けしたような色味が気に入っています。オリジナルの染料の関係なのか、スノーライオンってどれも褪せた感じなんですよ。冷気の侵入を防ぐダブルチューブ製法で作られた本品は、ダウンの量も凄い(カタログには"歩行用寝袋"と)。金子的エクスペディションダウン選手権の殿堂入りモデルです。

Snow Lion, a brand established in Barclay, designed this item. I love the sun-faded-like color. I'm not sure if it is because of the garment dyes they used, but all Snow Lion jackets come in faded-like colors. Made with the double tube method to prevent cold air from entering, this jacket is filled with an amazing amount of down (the catalog says it is a walkaround sleeping bag). One of the hall of famers of my best expedition jackets.

Snow Lion 16 BERING PARKA

アラスカ州ベーリング氷河から命名された
ベーリングパーカ。最終テストは実際に現
地で行われたらしい。これもダウンがパン
パンに詰まったエクスペディションな仕様
ですが、このモデルからバッフル構造が進
化し、より温かくアップデートしています。
例によって、カラーのオレンジ色も褪せて
いて、グラデーションのようなムラがある。
でも、やっぱりそこが好きなんですねぇ。

The name, Bering Parka, derives from the
Bering glacier in Alaska. The final product test
was said to take place there. The jacket was also
designed for extreme conditions and stuffed
with plenty of down. This model actually feels
even warmer with the advanced baffle
structure. The orange color is faded as usual
with uneven ombre, but I do like that.

● EXPEDITION PARKAS of U.S.A.

SIERRA DESIGNS × BEAMS PLUS   17   MINARET PARKA

MINARET PARKA

For really bitter conditions, the mountaineer relies on a "big downie" like our Minaret. We've made many improvements in this parka since introducing it last season, and we feel confident it will satisfy the perfectionist. The Minaret has double-quilt construction, with off-set seams. Fabric hidden between quilts is 1 oz. nylon ripstop; outer shell is 1.5 oz. ripstop. Average fill is 15 oz. of Silver Gray goosedown with average thickness of 2½". Parka is cut long; bottom comes below hips and has a draw closure. Nylon zipper front (#7) with down-filled, snap-fastened underflap. Cuffs are elastic-and-snap combination. Parka has attached, full hood with drawcord closure. Large front pockets are down-filled and close with Velcro tape. Behind these pockets is a second set, used as hand-warmers. Sizes: SM., MED., LGE., X-LGE
COLORS: Bright Blue or Red.
Avg. Wgt., 2 lbs. 2 oz. $70.00

DOWN PANTS

1.9 oz. ripstop shell filled with goosedown. Quilt-through construction. Zippers set on outside of legs allow pants to go on over boots. Adjustable waist, two patch pockets. Stuff sack included. COLORS: Blue or Orange. Wgt 18 oz. Provide inseam, outseam and waist measurements. $26.50

WIND PANTS

1.9 oz. ripstop fabric. Adjustable waist, two patch pockets and Velcro tape closure at ankle. Give waist, inseam and outseam measurements. COLORS: Blue or Orange. Wgt. 8 oz. $16.50

17

マウンテンパーカで有名なシエラデザインズのブランド最初期である1968年に作られていたダウンジャケットです。フロントジップがムキ出しでダウン量も含めてまさに着る寝袋。珍しい仕様ですね。ビームス プラス20周年の折に、米国製でダブルネーム復刻できたのはもはや奇跡！ 雑誌のフリマ企画でオリジナルを手放したことをずっと後悔しています。

Sierra Designs is a brand known for its mountain parkas, and this down jacket was made in the very early period of their history, 1968. With the exposed front zipper and plenty of down filling, it is truly a wearable sleeping bag. I think the design is quite unique. To commemorate the 20th anniversary of BEAMS PLUS, we reproduced this jacket in collaboration with the brand. It was fully made in the U.S. I used to own the Sierra Designs' original piece but had let it go for a flea market campaign of a magazine. I do regret that.

SIERRA DESIGNS × KAPTAIN SUNSHINE × BEAMS PLUS

18 MINARET PARKA

こちらは1970年のモデルをベースに、キャプテン サンシャインとビームス プラスがトリプルネームで製作したもの。キャプテン サンシャインのデザイナー児島氏が所有しているアーカイブを基に、当時と同じダブルチューブ製法をオリジナルファブリックで作り込みました。ビームス プラスの名作です。

Based on the original model released in 1970, this parka was made in collaboration with Sierra Designs, Kaptain Sunshine and BEAMS PLUS. We adopted the double tube method they used back then, referring the original piece owned by the designer of Kaptain Sunshine, Mr. Kojima. It's definitely one of the masterpieces of BEAMS PLUS.

● EXPEDITION PARKAS of U.S.A.

Recreational Equipment, Inc.   19 SUMMIT PARKA

こちらは登山向けに設計された超軽量モデル。カタログには"コンパクトさと軽量さが登山者におすすめ"との記載があります。薄手のシャイニーな生地は、ヨーロッパを彷彿させるもの。デザインモチーフもフレンチブランドのそれでしょう。ジッパーを使用せず、深い打ち合わせと大量のボタンで冷気をシャットアウト。

This super lightweight model was designed for climbing. The catalog says the item is ideal for climbers with the compactness and lightweight properties. The thin, glossy fabric has a hint of European designs. The detailing is also similar to the one seen on French brands' products. The front closure doesn't feature a zipper but a lot of buttons to prevent cold air from entering.

Recreational Equipment, Inc.   20   JAMET DOWN PARKA

冷気の侵入を防ぐ深い打ち合わせが印象的なこちらも、フレンチテイスト強めの一着。カタログを見ると、フランス製のモデルも実際にあったようですね。ボディに比べて極端に小ぶりなフードは、スタンドカラーの上部のジップから出し入れできる作り。これもヨーロッパものに多い仕様で、アンバランス感がなんとも愛おしいですね。

This parka with deeply overlapped front flaps shows the influence of French designs. According to the catalog, some of the brand's products were actually made in France. The hood looks small comparing to the shell and can be stored in the collar with a zipped closure. Many jackets made in European countries feature this detail. I do love the unbalance in size.

● EXPEDITION PARKAS of U.S.A.

Recreational Equipment, Inc. 21 DENALI EXPEDITION PARKA

生協組織に端を発するシアトルのアウトドアブランド、REIのアルパインシリーズ最高峰モデルです。ダウンぎっしりの迫力満点シルエットのボディは、サイドシームのない丸胴仕様で、ダブルチューブ製法。包み込まれているような着心地が病みつきになります。とにかく冷気を入れたくないという意思が伝わる作りで、フロントはダブルスナップとジッパー使いの豪華版。

REI is a Seattle-based outdoor wear brand that was originally established as a co-op organization. This bulky jacket is the prestigious model of their Alpine series. The down-filled shell is designed without side seams and manufactured with the double tube method. The fabric neatly wraps around your body and feels just comfortable. The dual press stud/zipped closure vividly reflect the brand's intention not to let cold air come in.

Recreational Equipment, Inc.
22 EXPEDITION BIVOUAC PANTS

左頁のデナリエクスペディションパーカの組下となるパンツで、野営時にはお腹から足先まで包む寝袋に早変わり。パーカを着てこれを穿けば、必要最小限の荷物で済むというシロモノです。ダウン量も多く、温かさは折り紙付き。パターンがよく、パンツ時のフィット感も良好ですし、ジップによる変身機構を含め、よく考えたなと感心します。

Matching pants to the Denali Expedition Parka shown on the left page. It turns into a sleeping bag that covers from the belly to toe during camping. By wearing these pants and the matching parka, you can travel with minimum baggage. These pants fits well and comfortably warms you up with plenty of down filling. Along with the way it turns its shape with zippers, all the details of this item are well thought out.

● EXPEDITION PARKAS of U.S.A.

Alpine Designs 23 EXPEDITION PARKA

バックパックの名声で知られる、コロラド発ブランドの一着です。マイナス60℃でも大丈夫と謳っていますが、さすがにちょっと眉唾ですかね!? さておき、オレンジのシェル×ブラックのテープの組み合わせは珍しく、カッコいい配色かと。それから、細かいところだとボタンがザラ目っぽいのも特徴で、イイ味出してるんですよ。

This jacket is made by Alpine Designs, a Colorado-based brand known for its backpacks. Though it is touted to withstand in a -60℃ environment, I find it a little hard to believe, to be honest. But the color combination of the orange shell and black tapes looks unique and cool. As for the detailing, I like the slightly grained surfaces of the buttons.

Frostline Kits　24　TUNDRA JACKET

完成品でなくDIYキットとして製品を販売する手法で成功を収めたフロストラインの、70年代に大ヒットを記録したモデルです。この一着もキットを購入した誰かが作ったんでしょうね。ダウンはふかふか。パネル数が少なく、全体的に簡易な作りなんですが、その分ボリュームが強調されるのか、不思議と魅力的なシルエットを描くんですよ。

Frostline Kits achieved success as a brand producing sew-it-yourself kits, and this model became a big hit in the 1970s. So, mine was made by someone who bought the kit. The down filling is so fluffy. The construction is simple with fewer parts, but it makes the silhouette look rather voluminous and charming.

●EXPEDITION PARKAS of U.S.A.

BEAMS PLUS 25 EXPEDITION PARKAS

ビームス プラスでは、クラシックなダウンジャケットを毎シーズン展開しています。これらは歴代のモデル。それぞれベースとなる型が存在しますが、より軽量な生地を製作したり、ダウンのフィルパワーを上げたり、必ずアップデートを加えるようにしています。あくまで"現代のクラシック"を提案したいので。手持ちのヴィンテージを上手く製品に落とし込めたときは、最高の気分！ 集めた甲斐があるってモンです。

BEAMS PLUS releases classical down jackets every season, and the jackets shown here are the ones we have made so far. Though each piece was designed based on a vintage jacket, we always add some improvements by using lighter fabric or offering a higher fillpower. Jackets we make have to be "modern classics". I felt so excited when I could make a satisfying product designed based on a vintage jacket in my collection, because it made all my effort worthwhile!

● EXPEDITION PARKAS of U.S.A.

Adventure 16 26 WIDE TEMPERATURE RANGE PARKA

サンディエゴの伝説的なアウトドアショップによって製作されたモデルです。肩周りはGORE-TEX®素材で切り替えされ、脇下にはジップのベンチレーションを装備。フード収納型のネックは4.5インチもの高さがあり、フロントを留めるベルクロはダイヤ型——と、凝った作りはローカル店の製品とは思えないほど。もうアメイジング！の一言です。

This jacket was made by Adventure 16, a legendary outdoor shop in San Diego. It features GORE-TEX® fabric around the shoulders and zipped ventilations under the arms. From the 4.5 inch high collar with a hood storage to the diamond-shaped Velcro, all the details are well-designed. I can't believe that a local shop made it by themselves. It's just amazing.

WILDERNESS EXPERIENCE 27 ULTIMATE PARKA

1973年創業のカリフォルニアのバックパック名門による本品は、エベレスト登頂を想定したスペック。80年代の製品で、生地がGORE-TEX®仕様になっています。擦れやすい袖に補強の当て布を添えた、タフな作りも持ち味。芯が入っているのか少々硬いフードのカタチや、ショートレングスなバランスも気に入っています。

Produced in the 1980s by Wilderness Experience, a prestigious backpack brand in California, this jacket features enough functions to summit Mt. Everest. The GORE-TEX® fabric is mainly used, while the arms are reinforced with patches to prevent abrasion. I also love the shape of the stiff interlined hood as well as the shorter length of body.

"パンパン"につまった夢を語ろう

金子　中田さんはいつ頃からアウトドアファッションにはまったんですか？

中田　いつだろう、おそらく中学3年から高校1年ぐらいが最初の僕のブームだったかな。

金子　そんなに早くからですか。

中田　そのときはアウトドアをファッションにしようみたいな雑誌の流れがあって。コロンビアのフィッシングベストとか、グラミチのショーツとか。グレゴリーのバックパックやマウンテンスミスのバッグも雑誌に載ってたと思います。1993年くらい。僕はファッション好きが集まる高校に通っていたので、皆でパタゴニアのメールオーダーをして、シンチラのスナップTを買って、アウトドアっぽいことしてるなぁって感動しましたね(笑)。

金子　そうなんですか！面白いですね。

中田　パタゴニアはすべてにおいて魅力的で。田舎者でしたので知識が無かったから、アウトドアウェアを買うなら、目白のパタゴニアストアに行くか、自由が丘のエルエルビーンへいくか、という感じでしたね。東京へ買い物へ行くときは自由が丘を回ってから渋谷原宿の古着屋を巡って、目白へ行く、というのが流れで。

金子　当時って、アウトドアものだけを集めている古着屋はありましたか？

中田　なかったんじゃないかと思います。ザ・ヴィンテージブームだったからアウトドアだけ、という古着屋はあんまり。フリマが一番のアウトドアウェアを沢山見れる場所だったかも。その後、大学2年か3年ぐらいのときに、ビームスのアウトドアスタイルに出会ったんです。デザイナーズものとかいろんなのをミックスするスタイルが衝撃的で。

金子　当時、エクスペディションダウンジャケットを着ている人っていましたか？

中田　それが記憶にあまりなくて。今よりも冬が寒かった気がするけれど、そんなに着てる人はいなかったような気がする。

金子　なんだか意外ですね。

Kaneko: From when have you been hooked on outdoor fashion?
Nakada: I'm not sure, but I first became crazy about it when I was 15 or 16.
Kaneko: Wow, that's quite early.
Nakada: Magazines at that time encouraged to make outdoor gear fashion, featuring items like Columbia's fishing vest and Gramicci's shorts. I think they picked up backpacks of Gregory and Mountain Smith, too. It was around 1993. Many of my high school friends were fashion enthusiasts, so we together mail-ordered Patagonia's clothes. I bought their Synchilla Snap-T and appreciated the fact that I did such an outdoorsy thing.
Kaneko: That's interesting.
Nakada: Patagonia was so fascinating in all. Because I was living in a rural area and not well-informed, I would go to either the Patagonia shop in Mejiro or the L.L.Bean shop in Jiyugaoka to look for outdoor wear. When I went to Tokyo for shopping, I always went to Jiyugaoka first and then visited vintage clothing shops in Shibuya and Harajuku before going to Mejiro.
Kaneko: Were there any vintage shops specialized in outdoor clothes at that time?
Nakada: I don't think so, they didn't carry many outdoor garments because typical vintage clothing was trending. I believe flea markets had the most selections of them. When I was a second- or third-year college student, I got to know the outdoor style of BEAMS. They mixed and matched various things such as designers' pieces. I was so shocked.
Kaneko: Were there anyone wearing an expedition down jacket around that time?
Nakada: I don't really remember. I think the winter at that time was colder than it is now, but few people wore it.
Kaneko: That's surprising.

● SPECIAL TALK with SHINSUKE NAKADA

28 EXPEDITION DOWN PARKAS

中田　でもビームスだと今は無きレーベル、B・E(ビームス エレメント)のスタッフは着ていたはず。マムートのすごいやつとか。あとはバランドレーっていうフランスのブランドのも着ていたと思います。

金子　バランドレー!?　知りませんでした。エクスペディションギアは、当時最先端の人たちが着ていたんですか？

中田　先端というか本当にコアな人たち、という感じで。ファッションとして意識され始めたのは、僕がビームス プラスのバイヤーをやっていた時代だと思います。2010年くらい。

金子　最初に作ったのはどのモデルですか？

中田　児島さん(現キャプテンサンシャイン デザイナー)と作ったパンパンのやつですね。

金子　ああ！　覚えてます。

中田　グレン・デニーの写真集(※1)を参考にして。60〜70年代当時の最先端の登山家たちを撮った写真集なんですが、そこにめちゃめちゃパンパンのダウンジャケットを着ている人が写っていて。ローファー履いて休んでいるんですが、パンツは60年代らしいスリムシルエット。アイビーなスタイルにダウンジャケットを合わせるのがほんとカッコよかったんですよ。児島さん私物のヴィンテージダウンを参考にして、カタチにした記憶があります。

金子　当時、ショップスタッフとして働いていましたが、これ2色買いましたもん。

中田　その後、集大成的にダブルチューブのやつを作りましたね。その時はシエラに依頼して作ってもらいました。2016年かな。

金子　ほかに思い入れのあるダウンジャケットはありますか？　アンライクリーの1stシーズンに出した短丈の一着はどうですか？

中田　それなら今季もやっていますよ。エンジ色の(※2)。これは僕もお気に入り。70年代の名作クラシックダウンをベースに、短丈にしたらどうだろう？というイメージで作った一着で。もちろんそれだけじゃなくて、ポケ

Nakada: Ah but I think some staff at B・E (BEAMS ELEMENT) that BEAMS expanded in the past were wearing it. Something like a big one from Mammut. Or the ones from a French brand called Valandre.

Kaneko: Valandre!? I didn't know the brand. Expedition clothes were worn by trendsetters at that time?

Nakada: Not really trendsetters but hardcore fans. Those clothes began to be recognized as a part of fashion when I worked as a buyer at BEAMS. It was around 2010.

Kaneko: What was the first jacket you made?

Nakada: The puffy one I made with Mr. Kojima (the current designer of Kaptain Sunshine).

Kaneko: Ah! I remember that.

Nakada: We referred Glen Denny's book (*1) that consists of the photos of the cutting-edge mountaineers in the 1960s and 1970s. In the book, there is a photo of a man resting in a super puffy jacket and a pair of loafers. Because the photo was taken in the 1960s, he wears narrow pants. Matching a down jacket with an Ivy League outfit looked really cool, so we made a similar jacket by referring a vintage down jacket that Mr. Kojima owned.

Kaneko: I was working at a BEAMS store as a salesperson around that time and got two of those in different colors.

Nakada: We later made another one with a double tube construction as a culmination. We asked Sierra Designs to manufacture it. I think it was in 2016.

Kaneko: Do you have any other down jackets that stay in your memory? How about the short-length one you designed for the 1st collection of Unlikely?

Nakada: We made it for this season, too. It's in maroon (*2) and my favorite. I first got an idea to shorten the length of a renowned down jacket made in the 1970s. And I also incorporate a pocket layout and a variety of other elements

46

ットワークだったり、いろいろなものを混ぜてデザインしてます。ポケットのサイズ感を誇張してハンドウォーマーにしたりとか。かなりアレンジはしていますね。

金子　中田さんってデザイナーの感性が強いですよね。たとえモチーフがあったとしても、原型があまり見えてこないんで。クリエーション魂を感じます。

中田　いやいや、同じ土俵にいても、そういうのをちゃんと研究してやっている人たちには勝てないですから(笑)。フィット感についても自分が着たいバランスに変えてます。ただこのダウンジャケットも、テープで挟み込まずにドットボタンを付けたり、グレーのテープを使ったりとか、昔の方程式にはなるべく沿って作っていますね。一方で素材は、昔ながらのコットンナイロンだと水が入っちゃうんで、ナイロン100%なんだけど昔風に見えるのを探したりして。裏地もイタリアのリモンタ社のものを採用しています。外はクラシック、中はラグジュアリーみたいな。

金子　なるほど、まさしくアレンジされていますよね。ほかに中田さんがダウンジャケットを作るときのポイントはありますか？

中田　パンパン具合はあるかもしれない。超ボリューミーなシルエットが好きで。あとフードのボリューム感にもこだわりますね。

金子　それは大きく作るってことですか？

中田　そうだね。見え方もあるし、やっぱり防寒としての機能。あとは電車で寝るときのクッションとして気持ちいいか、とか(笑)。

金子　中田さんは通勤時間も長いから(笑)。エクスペディションダウンベスト(※3)の話も聞きたいです。ベストなのにエクスペディション仕様、というのが面白くて。

中田　なかなかないでしょう？　去年作ったのですが、私物のクラシックなダウンジャケットをベースにベストにして、パンパンに膨らませて。で、それにさらにライナーを付けて。ダウンベストとしても着られる3way

into my designs. Some of the pockets are made big in size to be used as hand warmers. I have put my own spin on it.
Kaneko: The way you think is more like a designer. It's not possible to detect your inspiration source from garments you made. You've got a creative spirit.
Nakada: No, no. There are people who design clothes based on elaborate researches. My expertise is way less than theirs though, even if we are in the same field. When designing, I also change the fit of the original piece according to my preference, but I try to follow the original design and methods as much as possible. As for this jacket, press studs are attached without being inserted in a tape, while the tape is in gray color. The fabric is, however, 100% nylon, although the original one is made of water-sensitive cotton nylon. So I looked for fabric that looks like the one used on vintage garments. The lining fabric is made by Limonta in Italy. It's like, the jacket is classic outside and luxury inside.
Kaneko: I see, there is your spin on it. What is the thing you focus on when designing a down jacket?
Nakada: A puffy feel, I guess. I prefer a super puffy silhouette. And the hood has to be voluminous, too.
Kaneko: Do you mean it has to be big?
Nakada: Yeah. I do care about the appearance and the cold protection. I'm also particular about how comfortable it is as a cushion when I sleep on the train.
Kaneko: That's because your commute is long. Could you tell me more about your expedition down vest (*3)? It sounds interesting that a vest comes with expedition-quality detailing.
Nakada: Quite unusual, isn't it? I designed it last year based on a classical down jacket I own. I made it super puffy. This is the updated version with lining and in 3 way design. You can wear it also as a vest.

● SPECIAL TALK with SHINSUKE NAKADA

28 EXPEDITION DOWN PARKAS

アップデートしたんですよ。アウター側はダウンでなく、ポリエステル中綿を詰めています。ダウンって温かいけれど、濡れちゃうと復元しにくいから、雨風にさらされるアウター側は中綿にして、シェルにも超撥水の素材を使って。

金子　なるほど。それにしてもダブルで着るとすごいボリュームですね。まさにエクスペディション。マッチョになります。ほかにこんなダウンジャケットを作ってみたいというのはありますか？

中田　次は、これまたクラシックな見た目の立ち襟にフードを収納したこれ(※4)をモチーフに作りたいと思っていて。上にジップが付いていて、見映えがイイなぁと。

金子　いいですね！

中田　ツヤッとした素材が特徴的なんだけど、70〜80年代ぐらいのスキー用ダウンジャケットって大体こんなタイプで。モンクレールもそうだしヒマスポーツのもそう。色は大体が赤と青のコンビネーションですよね。

金子　今日は中田さんとディテールの話がしたかったのでコレクションの中からいくつか見繕ってきました。中でも、ディテールが一番好きなのはアドベンチャー16の一着(※5)ですね。フロントが菱形のベルクロで。

中田　イイなぁ〜、アドベンチャー16、好きです。じつは一番最初にロサンゼルス出張行ったときに一番感動した店が、ここでした。

金子　本当ですか！　もうないですもんね。

中田　じつは最初に連れてってもらったアウトドアショップがここだった。2003年とか2004年頃かな。新品も古着も売っていて。

金子　僕も行ってみたかったなぁ。

中田　僕はアドベンチャー16のポーチ(※6)を買っていまだに使っていますよ。仕切られていて、いろんな国に行く出張のときにも便利。

金子　トリプルジッパーなのがいいですね。

中田　赤い方のダウンジャケット(※7)は？

金子　ポイントファイブの一着なんですけど、

The outer vest is insulated with polyester filling because it is exposed to wind and rain. Down filling is warm but doesn't go well with water. So the shell is made of highly water repellent fabric.
Kaneko: Oh I see. It's really voluminous when I wear both of them. Truly for expedition. It makes me look muscular. Do you have any particular down jacket you want to design?
Nakada: I want to design based on this jacket (*4). It comes with a stand-up collar with a hood stored in it. There is a zipper on it and the overall design looks nice.
Kaneko: That's great!
Nakada: The glossy fabric is characteristic. Ski down jackets made between the 1970s and 1980s are mostly made of a similar type of fabric. Moncler and Himasports used it, too. As for the color, many of them are in the combination of red and blue.
Kaneko: I picked up some jackets from my collection today to talk about the detailing with you. My favorite is this one (*5) from Adventure 16. The front closure features diamond-shaped Velcro.
Nakada: Looks great. I like Adventure 16. It was actually the most impressive shop I'd visited during my first business trip to Los Angeles.
Kaneko: Really! They are no longer in business.
Nakada: It was the first outdoor shop I was introduced then. I think it was in 2003 or 2004. They carried both new and used products.
Kaneko: I wanted to visit, too.
Nakada: I got Adventure 16's pouch (*6) and am still using it. With a neatly divided interior, it's so handy when you travel to various countries for business.
Kaneko: The triple-zipper system looks nice.
Nakada: How about the red down jacket (*7)?
Kaneko: It is from Pointfive and characterized by the diamond-shaped down packs and dual press studs. What

48

菱形のダウンパックやダブルスナップが特徴的で。面白いのが、着脱できる汚れ防止の襟カバーが付いているんですよ。

中田 おお、洗えるんだ！ たしかに首周りが汚れるのすごいイヤだよなぁ。

金子 肩がドルマンスリーブみたいになっているんですが、これは肩を上げやすいような構造と当時のカタログに書いてあります。そのための菱形のキルトなんですよね。

中田 へぇ、勉強になりますね。

金子 中田さんが好きなダウンジャケットの着こなし方を教えていただけますか？

中田 僕は重ね着が好きなので、スウェットやニットを着て、その上に布帛のブルゾン系のものを着て、ダウンジャケットを着るのがセオリーですね。前を開けると、素材、色、長さのグラデーションを楽しめる。謳ってはいないんだけど、これが僕のヘビーデューティーの受け止め方なんです。機能面でいえばECWCS(Extended Cold Weather Clothing System)の発想にも似ていて、重ね着が一番、理に適っているなと。茂はどう？

金子 自分はなるべくクラシックアウトドアを感じる原色系のダウンジャケットを着て、色と色を合わせるのが気分です。

中田 今のビームス プラスチームは凄いと思います。色使いがぶっ飛んでて素晴らしい。

金子 柄物と色物が戦ってる感じですね(笑)。でも、気持ちいい色合わせってあるじゃないですか。その色合わせの中で、さらに柄を組み合わせるというのが個人的には好きな着こなし方かなと。今のビームス プラスは、中田さんのディレクター時代から変わったと思いますか？

中田 もう全然違いますよね。先輩たちから受け継いだバトンを、溝端(ビームス プラス ディレクター)と茂が昇華させてくれたという気がします。

金子 昇華って(笑)。全部先輩たちのおかげです。でも、ありがとうございます！

I like about it is that it has a detachable cloth on the collar to prevent dirt.

Nakada: Oh this one is washable! Yeah, it's really annoying that the collar gets dirty.

Kaneko: The sleeves look like dolman sleeves. The catalog published back then says it makes the wearer easy to raise the arms. And that is also why the down packs are designed in a diamond shape.

Nakada: Wow, that was enlightening.

Kaneko: What is your favorite way to style a down jacket?

Nakada: Because I like layered looks, I always layer a cotton blouson or jacket on a sweat shirt or sweater before wearing a down jacket. So that you can show the contrast of different fabrics, colors and length with your jacket opened. I don't officially say it but that is what I think the heavy-duty style is. Just like the ECWCS (Extended Cold Weather Clothing System) of the U.S. military, I think layered style makes perfect sense in terms of functionality. How about you?

Kaneko: I currently like to wear a down jacket in a primary color as it has a classical outdoor feel. I would then match it with another vivid colored item.

Nakada: The current BEAMS PLUS team is so great. The way they use colors is just eccentric in a good way.

Kaneko: There are so many colors and patterns. But we may have some combinations of colors that make us comfortable. I prefer adding a pattern to one of the combinations. Is the current BEAMS PLUS team different from when you served as the director?

Nakada: It's completely different. Mizobata (BEAMS PLUS Director) and you have elevated what we had passed down to them to the next level.

Kaneko: It's thanks to you and other predecessors, but I'm happy to hear that.

EXPEDITION PARKAS OF EUROPE

究極の防寒具たるダウンジャケット
ルーツを辿ればここに行き着く

アメリカのアルパインダウンジャケットを調べるほど、そのルーツはヨーロッパにあることがわかります。さまざまなブランドが初期に採用していたダブルスナップの意匠も、やはりヨーロッパ由来。フード収納型のスタンドカラーもしかりで、そんなディテールを面白く思ううちに、気づけばヨーロッパのエクスペディションダウンジャケットもコレクションしていました。シャイニーな生地や配色の妙も、ヨーロッパのそれの面白さです。写真のダウンジャケットは、イギリスのヴィンテージダウンをベースに"現代のクラシック"へとアップデートして製作した、パレス スケートボーディング × ビームス プラスのコラボレーションモデル。英国気分を演出すべくパッチワークニットを着込み、足元にはマウンテンブーツを連想させる佇まいのクラークスを合わせました。

The origin of alpine down jackets,
the ultimate winter garment

As you learn more about American alpine down jackets, you may realize that they originate in Europe. For example, dual press stud closure, a type of front closure adopted by many American brands in the early period, has its roots in Europe, as well as a stand-up collar with a hood storage. I started collecting European expedition jackets, too, because such detailing fascinated me. The use of glossy fabric and unique color combinations also characterize European down jackets. The jacket I wear in the photo was made in collaboration with Palace Skateboards and BEAMS PLUS by elaborating the design, which is based on a vintage piece from a British brand, into a "modern classic." A patchwork sweater and Clarks' mountain-ish boots add a touch of Britishness to the look.

● EXPEDITION PARKAS of EUROPE

Blacks 29 DUVET JACKET

テントやギアも展開していたイギリス老舗のダウンコレクション。表地にコットンを用いた、50年代の古いものもあります。僕が着用したジャケットの表地はナイロンで、ダブルスナップや淡い発色にヨーロッパの趣がむんむん。リブの袖口やポケットレスの作りも、よく見られる仕様です。寝袋で寝る時は、ボタンを外側に留め、腕を袖から抜いて身頃側に入れるのがセオリー。

Here is my collection of down jackets made by Blacks, an established British manufacturer which has also produced tents and other outdoor gears. There is one made in the 1950s with a cotton outer shell, though the one I wear on this page is made of nylon. The dual press stud closure and pale color boost a European mood. When sleeping in a sleeping bag, you should widen the width of the dual press stud closure with the sleeves tucked inside.

● EXPEDITION PARKAS of EUROPE

MONCLER 30 DUVET JACKET

アメリカブランドがこぞってモチーフにした、ダウンの雄・モンクレール。1952年に創業し、フランス人で初めてヒマラヤ登頂に成功した登山家リオネル・テレーをアドバイザーに迎え、本格的なダウンウェアの開発に着手しました。こちらの60年代のモデルには氏の名前入りのタグが付いており、ダブルスナップも象徴的。別のジャケットのフードを付けてバイカラーを楽しむのも今の気分です。

American brands have all taken inspiration from the iconic down brand, Moncler. Founded in 1952, the company brought on Lionel Terray, the first French mountaineer to successfully climb the Himalayas, as an advisor and began developing high-performance downwear. This model from the 1960s also features a tag with his name and the brand's signature double-snap design. Adding a different hood to enjoy a bi-color look is a stylish choice for today.

MONCLER 31 KARAKORUM

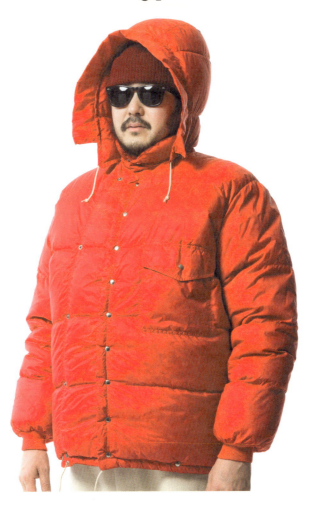

ブランドのアイコンモデル。1954年にイタリアのカラコルム登頂隊のオフィシャルスポンサーに選ばれ製作されたモデルです。左頁のジャケットと同様のデザインですが、1つ大きな違いがあります。それは左胸にフラップポケットが付いていること。ヨーロッパのダウンジャケットは雪の浸入を考慮してポケットのないものが多いのですが、これは例外。本デザインもテレー氏の功績のひとつです。スカーレットカラーを着用した当時の広告はいつ見ても刺激を受けます。

The brand's iconic model, Karakorum. This jacket was created in 1954 when Moncler was selected as the official sponsor for the Italian Karakoram expedition team. It features the same design as the jacket on the left page, but with one major difference—a flap pocket on the left chest. While many European down jackets typically lack pockets, this is an exception and stands as one of Lionel Terray's design contributions. The vintage advertisement featuring this jacket in scarlet red remains an inspiration to me to this day.

● EXPEDITION PARKAS of EUROPE

Pointfive 32 DUVET JACKET

テッカテカのナイロンに、腕の可動域を高めるためとされるダイヤ型のキルトパターンが、もうたまらない！ イギリス老舗による70年代に作られたジャケットです。フードが欠損していますが、取り外せるネックカバーが付いていて襟が汚れないようになっていたり、参考になるディテールばかり。今一番気分な一着ですね。

I just love the glossy nylon fabric and diamond stitches (allegedly designed for better mobility) of this jacket! It is made in the 1970s by Pointfive, an established British brand. Although the hood is missing, it comes with so many features to refer to such as detachable neck cover to avoid dirt around the collar. This is my latest favorite.

Pointfive 33 DUVET JACKET

こちらもポイントファイブのモデル。素材も仕様も左頁のモデルとはまったく異なりますが、これもお気に入りです。配色が最高じゃないですか！ ホントは赤がオモテ、水色がウラ。でも裏返して着ると、水色の生地と赤のリブ袖の対比がイイ感じになるんです。さらに所有者の名前がステンシルで入っていて、物語を感じるところもヴィンテージの面白さ。

This one is also made by Pointfive. I love this, too, though the fabric and design are both totally different from the one on the left page. The color combination is just fabulous! The red color should be on the face side, but if you wear it on the flip side, the pale blue fabric looks gorgeous with the accent of red ribbed cuffs. I also like that the name of the original owner is stenciled and tells a bit of story of the person. Such a thing is seen only on vintage clothes.

● EXPEDITION PARKAS of EUROPE

MOUNTAIN EQUIPMENT 34 CERRO TORRE

イギリスの老舗より、ウルトラシャイニーなオレンジと裏地のブルーの対比が魅力的な70年代のジャケットです。このブランドのジャケットって逐一配色がイイんですよね。セロ・トーレの名前は、パタゴニアの名峰から。配色はもちろん、生地やシルエットまで参考になるポイントが詰まった一着です。

Mountain Equipment, an established British manufacturer, made this jacket in the 1970s. The super glossy orange surface fabric makes a great contrast with the blue lining. Jackets made by the brand always come with a beautiful color combination. The name, Cerro Torre, derives from the renowned mountain of the same name in Patagonia. The color scheme, fabric, and silhouette all make this piece my great source of inspiration.

Himalchuli 35 DUVET JACKET

フランスのマウンテンイクィップメントといえば、ヒマルチュリ。その名の由来は、ヒマラヤ山脈の一峰からきています。特徴的なコルセットのようなフード収納型のネックディテールはブランドのアイコン。ポケットワークとブラックのナイロンキャンバスの補強布で、このジャケットの機能美が完成します。

When talking about of french mountain equipment brand, Himalchuli is the brand. The brand's name originates from the mountain forming a part of the Himalayas. The iconic corset-like collar is designed with a hood storage. The pockets and black nylon canvas reinforcing fabric complete the jacket's functional aesthetics.

FIELD JACKETS

デザインに、ディテールに、素材に
特別な目的をなすための理由が必ずある

フィールドジャケット。字義どおりなら屋外で着るジャケットとなり
ますが、野良仕事からハンティングやフィッシング、スキーといった
アウトドアスポーツにいたるまで、多様な目的に即してデザインされ
た無数のそれが存在するのです。これらもまたエクスペディションな
ギア。用途に特化したカタチやディテールは、なぜこうなのか？とい
う僕の想像力を掻き立てます。そして広大な敷地をもつアメリカなら
ではの多様性が、魅力を無限に押し広げてくれるのです。僕が着てい
るのは、1964年のオーストリア冬季五輪で、USAチームのオフィシャ
ルウェアに採用されたスロープコート。スキームードに鑑み、雪を連
想させるホワイトのハーフジップニットを合わせてみました。

There is always a reason
for the design, detailing and fabric

The design of a field jacket varies according to the purpose it is made for. From farming
to hunting to fishing to outdoor sports such as skiing, there are so many types of field
jackets. And they are one of expedition garments, too. Design and detailing made for a
specific purpose always stirs up my imagination on what it is for. America's diverse
types of products have an inexhaustible charm. A slope coat I'm wearing was originally
made to outfit the U.S. national team for the 1964 Winter Olympics in Austria. To boost
the winter sports mood, I match it with a snowy white half-zipped sweater.

● FIELD JACKETS

L.L.Bean 36 BEAN'S DOWN MACKINAW

大好きなエルエルビーンより、トラディショナルなマッキノースタイルのジャケットにダウンを詰めた一着です。カッコいいんだかカッコ悪いんだか、ちょっぴり野暮ったいところに惹かれます。肩にもたっぷりダウンが入っているので、着るとマッチョ感が凄くて（笑）。屈強な男に似合う、じつにアメリカらしいアウターかと。

My favorite brand, L.L.Bean, made this traditional mackinaw-like down-filled jacket. It looks both stylish and ugly, yet I actually find the uncoolness rather charming. Because the shoulders of the jacket are also filled with plenty of down, it buffs me up when I wear it. I think it suits a muscular guy. Such an American outerwear.

WOOLRICH 37 8PT BARRIER HUNTING COAT

こちらもアメリカを代表するアウトドアブランドの、いなたい70年代のダウンコート。レザーパイピングがなんともミスマッチで、これが野暮ったさに拍車を掛けている!? ともあれラギッドなムードや真っ赤な発色は、ウールリッチらしさ満点！60/40で軽い着心地なところもお気に入りのポイントです。

This rustic down coat is made in the 1970s by Woolrich, a renowned American outdoor wear brand. The leather piping doesn't really match with the design and probably spurs the uncouthness of this jacket. But the ruggedness and vivid red color convey the Woolrich mood! I like that it is made of 60/40 cloth and so light in weight.

● FIELD JACKETS

Alyeska 38 PIPELINE DOWN COAT

アリエスカは、アラスカを横断するパイプライン建設に従事する人に向け、東海岸アウトドアブランドが70年代半ばにOEM生産していたブランド。生地は難燃ノーメックス®。レザー補強がいい味を添えています。フードは当時のオーナーがファーから迷彩にカスタムした仕様。ナイスセンス！

Alyeska is an OEM brand by the outdoor brand from East Coast for construction workers of the Trans-Alaska pipeline in the mid 1970's. The fabric is NOMEX® and the leather patch adds a great accent to it. The previous owner seemed to change the hood fabric to camo cloth. I love his taste.

Patagonia 39 DOWN SWEATER

Yジョイントスリーブという、クライミング時でも腕を上げやすいように施したパターンが特徴的なダウンセーター。胸と腕部分はダウン量を増やしていて、保温力を高め冷えないようにデザインされています。レイヤードするベストも存在するのですが、ダウンシームの位置をズラしていて、冷気が入りこまないよう工夫している。そして、色使いが最高！

This down sweater features Y-joint sleeves, a type of sleeve design that allows for better mobility during climbing. The chest and arms are filled with extra down to improve heat-retaining properties to keep the areas warm. A matching underlayer vest was available and the position of seams of down pannels is misaligned with the one on the sweater to prevent cold air coming in. The color combination is brilliant, too!

● FIELD JACKETS

WOOLRICH 40 HUNT PARKA

ベーシックなハンティングコートのスタイルかと思いきや、ポケットはアシンメトリー。フラップ付きのカートリッジホルダーが目を惹く、ウールリッチらしい一着です。カモフラージュ柄も相まってアメリカンな雰囲気満点ですが、左右非対称なデザインがこうも魅力的に映るのは、そのほうが合理的だからなのか。違和感のあるデザインは大好物。

It looks like a simply designed hunting coat but actually features asymmetry pockets. The characteristic flapped cartridge holder enhances the Woolrich style, while the camo pattern exudes an American vibe. Why does the asymmetry design look so fascinating? Is it because it is more practical? Anyway, I love such oddity.

Marmot 41 THE PICKET PIN PARKA

アイテム名は、米国西部の山地に生息するリスの呼称に由来。冬のスキーや登山、山作業向けのモデルです。発色のよさもさることながら、当時オプションでオーダーができた胸の配色が最高で。大きな丸形と長方形のベルクロ使いもかっこいいですよね。フィット感のよさと動きやすさは、売りにしているだけあって間違いありません。

The name of this jacket derives from a nickname of a squirrel that lives on mountains in the western United States. Designed for skiing, winter mountaineering and forestry work, this jacket is characterized by the vivid colors. I love the stripe on the chest which could be added as an option back then, as well as the circle and rectangle Velcro. The selling point of this jacket is the fit and movability, and the quality is just as it is advertised.

● FIELD JACKETS

Gerry 42 SLOPE COAT

グローブをしたままでも操作しやすい、樹脂製のトライアングルトグルが特徴的なコートたち。1964年オーストリア冬季五輪USAチームのオフィシャルコートとして採用されていました。その後1965年にはブランドのインラインとしてリリースされ、カタログにも登場します。見出しには"GERRY MASTERPIECES"との表記が。ブランドにおけるスキーウェアラインが開花した瞬間ですね。

These coats are characterized by resin toggles designed for easy access even with gloved hands. It was originally supplied to the U.S. national team for the 1964 Winter Olympics in Austria as an official garment. In the following year, the manufacturer, Gerry, generally released the item and featured on its catalog with a headline that reads "GERRY MASTERPIECES". It was the moment when the brand's ski wear started its history.

White Stag 43 SKI JACKET

オレゴン州ポートランドで創設されたハーシュ・ワイス社のスキーレーベルより。スキーウェアらしくタイトなシルエットで、スタイリッシュな趣のジャケットです。ダウンパックが細かく、ダウン量は少なめ。動きやすさを重視した作りですね。ジップの引き手には、ブランドアイコンであるトナカイのチャームが付属。シックなグリーン色と併せて気に入っています。

White Stag is the name of a ski wear brand run by Hirsch Weis, a company established in Portland, Oregon. The tight shape makes the overall silhouette stylish, while the shell is divided into small packs moderately filled with down for better mobility. The icon of the brand, a small reindeer, is attached to the zipper pull. I love the accent as well as the chic green color.

● FIELD JACKETS

Early Winters 44 THE LOST WORLD PARKA

GORE-TEX®使いの先駆けブランドが手掛けた渾身の一着。特徴的なのはフードに施されたダブルドローストリングシステム。フードロを絞るコードに加え、額とフードを密着させるためのコードが備えつけられており、どこを向いても顔とフードがともに動きます。よって、視界を妨げないのです。大開口のベンチレーション機能も完璧。

Early Winters is one of the pioneers of the GORE-TEX® products. This masterpiece is characterized by the double drawstring system on the hood. One of the strings is designed to tighten the hood and the other to make the hood stick to the forehead. With the system, the hood keeps up with the movement wherever you move your face to, and never limit your visibility. There are also huge ventilations under the arms.

Levi's 45 DENIM MOUNTAIN PARKA

70〜80年代の初頭にかけ、アメリカでマウンテンパーカが流行っていたんだなぁというのを証明する一着。かのリーバイスも作っていたんです。タブやタッグボタンなど、ジーンズに使用されているパーツが盛りこまれていてブランドらしさ満点。デニムならではのユーズド感が他にはない雰囲気でたまらないんです。

Mountain Parkas became a trend in the U.S. from the 1970s to early 1980s, and this jacket proves it. Even Levi's made one. Some jeans-like details such as a tab and tuck buttons reflect the brand's aesthetics. The worn-out look of the denim is definitely one of a kind.

EDDIE BAUER ITEMS

エディ・バウアーがアメリカの
ダウンウェアの歴史を築いた

ダウンジャケットを世界に先駆けて生産したエディ・バウアーの歩みは、アメリカのダウンウェアの歴史でもあります。ハンティングやフィッシングギアとして生まれたダウンジャケットはその後、アメリカ空軍のフライトスーツにも姿を変え、アラスカ探検やヒマラヤ遠征といったエクスペディションな環境を生き抜くためのギアへと発展。その中心にはいつもエディ・バウアーがありました。ダウンメーカーの第一人者だからこそのアイテムバリエーションの豊富さは、まさに圧巻の一言です。というわけで、名作スカイライナーを皮切りに、ダウンキャップにダウンマフ、ダウンマスクにダウンカバーオールと、これでもか！というほどのダウンウェアを重ね着してみました。いやはや、あらためて凄いバリエーションですね。

Eddie Bauer,
the pioneer of american down clothing

The history of American down clothing began when Eddie Bauer started producing their down jacket ahead of the rest of the world. Although the jacket was originally invented for hunting and fishing, the U.S. Air Force later transformed it into a flight suit and the functionality had been continuously improved to be worn in Alaska, Himalayas and other extreme conditions. And, whenever major changes arose, Eddie Bauer played the key role. Their broad range of down products are just incredible. So, I layered them as many as possible, including their renowned Skyliner jacket, down cap, down muff, down mask and down coveralls. Now you see how incredible they are.

73

● EDDIE BAUER ITEMS

Eddie Bauer 46 DOWN ACCESSORIES

エディー・バウアーのダウン製品、その多さは、小物に象徴されます。とりわけユニークなのは、狩りや釣りの待ち時間に手を温めるダウンマフ。コヨーテカラーのそれはテストサンプルで、ジップポケット付きにアップデートされているよう。かなり希少ですね。赤のネックウォーマーとフード＆ディッキーは、アクセントにも重宝します。

Eddie Bauer has released so many types of down-filled products. One of the most unique things among them is down muffs, a cylindrical covering into which the hands are placed to keep them warm during hunting or fishing. The one in coyote color in the photo was originally made as a prototype and later updated with zipped pockets. Truly a rare piece. The neckwarmer and hood & dickey are ideal items to give a vivid accent to your look.

Eddie Bauer 47 DOWN WEARS

品のいい面構えのコートはシアトルのトラディショナルブランドとのダブルネーム。鳥柄プリントのダウンライナーがじつに洒落ています。コーデュロイ使いのインナーは、ジレ風のフロントがユニーク。寒冷地のワーカー用のツナギやジャンプスーツも珍しいですね。こうした意外なダウンウェアが充実するのも、その道のパイオニアならではでしょう。

The chic coat shown on the top was made in collaboration with a Seattle-based traditional brand. The bird pattern lining is amazingly stylish. The inner jacket on the under-middle comes in a unique design and looks as if layering a cord vest. The jumpsuit and coveralls for workers in cold climates are rare pieces, too. Such unusual down-filled garments could only be made by the pioneer, Eddie Bauer.

●SPECIAL TALK with MANABU HARADA

48 DOWN VESTS

Playing with down vests!

原田 学
Manabu Harada
Stylist

1972年生まれ。古着をミックスしたスタイリングの第一人者。ヴィンテージアイテムを紹介する数多くの書籍を手掛ける。自身の好きを感覚で紹介する書籍『THE SUKIMONO BOOK』は著者であり、原稿やイラストも担当。

Born in 1972. The pioneer of vintage mix-and-match style. He has been active in various media, including magazines, advertisements and TV. His book "THE SUKIMONO BOOK" introduces his favorite vintage pieces together with his texts and illustrations.

遊び方は無限大。型にはめないおもしろさ

<u>金子</u> 古着ミックスのスタイリストといえば原田さん。ずっと憧れていました。

<u>原田</u> いやいやぁ。古着をスタイリングする人がなかなかいないんでね。もう何年もやっていますが、古着屋さんのリース伝票の記録をみると、僕の前に借りた人はいつも僕。

<u>金子</u> 古着屋さんだと、繋がりがないと服も貸してくれなさそうですもんね。

<u>原田</u> 最近は貸さないところが多いですね。僕の場合はまあ、友達みたいなものだから。

<u>金子</u> 元々はそんなに古着がお好きなわけではなかったとインタビュー記事で読みました。

<u>原田</u> 好きな人だと学生のときから代々木のフリマに通い詰めて、みたいな話があると思いますが、僕はまったくそういうのがなくて。今の仕事を始めてからのめり込んだ感じです。25歳くらいのときからですかね。

<u>金子</u> その時代の古着屋さんって、怖い印象があります。

<u>原田</u> 洋服屋さん自体が怖かったですもんね。ご飯の時間になると営業時間でもドアの鍵を閉めてお客さん入れなくしちゃったり。

<u>金子</u> うわぁ、時代ですね(笑)。古着のなかでも、原田さんがアウトドアアイテムに興味をもったのはなぜだったのでしょう？

<u>原田</u> ある雑誌のスタイリングをしていたのですが、そこは編集長がおおらかで僕らスタイリストが面白いと思った企画は自由にやらせてもらえたんですね。2000年を過ぎた頃だったかな、神戸のジャンクショップへ行ったら、アウトドアブランドのバッグが大量にあって。グレゴリーなどの一部は人気で高値が付いているのに、ほかは軒並み手頃だった。様々なアウトドアブランドの様々なスタイルのバッグを紹介したら面白い？と、そういうページを作ったんです。すると、この年代の製品はディテールがどうとかこうとか興味がわいて、アウトドアの世界にどっぷりと。

<u>金子</u> 原田さんが出している『THE SUKIMONO BOOK 01』もバッグが主役でしたね。

Kaneko: Speaking of vintage mix-and-match style, you are the first person who comes to our mind. I've been admiring you.
Harada: That's just because there are few people who style looks with vintage clothes. I've been doing this for years but when I look at vintage clothing shops' records, I am always the person who borrowed clothes from them before me.
Kaneko: It seems difficult to borrow clothes from vintage clothing shops without connections.
Harada: Many of them don't lend things they sell recently. But I'm like a friend of them.
Kaneko: I have read an interview that says you were not so fascinated by vintage clothes at the beginning.
Harada: Enthusiastic people may have an episode, like, they have frequented a flea market in Yoyogi since they were students. I don't have such an episode at all. I got hooked on it after starting my career as a stylist. I was about 25 then.
Kaneko: I felt vintage clothing shops around the time were unwelcoming.
Harada: Clothing shops at that time were generally unwelcoming. They locked the door during lunchtime to shut out customers.
Kaneko: Ah, that was typical back then.
Harada: When working for a magazine as a stylist, the editor-in-chief was so generous that we stylists were given freedom to do whatever we thought interesting. I think it was a little after 2000 when I found lots of outdoor brands' bags at Junk Shop in Kobe. Some items, such as ones from Gregory, were popular and priced high, but all the rest were reasonably priced. So, I thought it might be fun to show various kinds of bags made by outdoor brands and edited an article with the idea. I then became intrigued by things like the age of manufacture and detailing, and got utterly absorbed in it.
Kaneko: Your book, "The Sukimono Book 01", also focuses mainly on bags.

● SPECIAL TALK with MANABU HARADA

48 DOWN VESTS

原田　アウトドアウェアの古着って、ザ・ノース・フェイスのダウンジャケットなんかが流行った90年代にはみんな着ていたけれど、専門的に掘って着ている人って、そんなにいなかった。エルエルビーンとかエディー・バウアーの古着人気が上がっているのって、最近じゃないですかね？
金子　ワークウェアの値段が上がり過ぎちゃって、アウトドアに流れてきているのかもしれないですね。でもパタゴニアの古着も最近、すごく値段が上がっていますよね。
原田　たしかに。日本に一番数が集まってるんじゃないかな、パタゴニア。間違いなくカッコいいし好きだけど、流行っているから僕は逆に封印しています(笑)。
金子　流行ってるからスタイリングに使いたくないなんてこと、あるんですか(笑)？
原田　というか全部そう。何年もベースボールキャップが流行っているじゃないですか？だからキャップはほとんど使わないですね。
金子　マジですか(笑)！
原田　ええ。トレンドを見せるだけじゃなくて、次に繋がることをやりたいと思っていて。だから流行は避けるようにしているんです。
金子　さすがですね。今日の原田さんもシャツの中にダウンベストを着るという、かなり個性的な着こなしですね。
原田　これは昔からやっていて。家で寒いときも、セーターとかあまり着ないんです。ロンTの上にダウンベストを着ることが多くて。で、外へ行くときはそのまま上からシャツを羽織る。車の運転もしやすいし機能的かなと。
金子　原田さんらしいスタイルですね。でも、この着方がもし流行ったら？
原田　そのときは中にシャツを着る(笑)。
金子　流行らないといいですね(笑)。ところで原田さん、ダウンベストはブランドで選ぶんですか？　デザインですか？
原田　デザインですね。ただトレイルワイズとか、イイものが多いマイナーなブランドも

Harada: When The North Face's down jackets became a trend in the 1990s, everybody wore vintage outdoor clothes, but few of them had deep knowledge about it. It's only recently that vintage clothing from L.L.Bean and Eddie Bauer has become popular, isn't it?
Kaneko: I guess because the price of vintage workwear has gone up too much, people are shifting to outdoor wear instead. But the prices of vintage Patagonia clothing have also increased significantly recently.
Harada: Indeed. I think Japan has the most of vintage Patagonia clothes. They are definitely cool and I like them, but I try not to use them when I style a look because they are trending.
Kaneko: You don't want to use it for a look because it's trending? How can that even happen?
Harada: Yeah, I'm always like that. Baseball caps have been a trend, right? So I seldom incorporate it in my look.
Kaneko: Seriously?
Harada: Yes. What I do is not just showing trends. I'd like to do what leads to the next. And that is why I avoid trends.
Kaneko: Amazing. Today you are wearing a down vest under a shirt. That's so unique.
Harada: I've actually been doing it for a while. I don't really wear a sweater or things like that even when I feel cold at home. I'd rather layer a down vest on a long-sleeve tee. To go out, I often put a shirt on it. I think the outfit is practical as it's easy to drive in it.
Kaneko: That's the look that suits you. What if this style became trending?
Harada: In that case, I'll wear a shirt underneath.
Kaneko: I hope it doesn't become popular. By the way, do you choose your down vest based on the brand? Or design?
Harada: Design. But there are some minor brands, such as Trail Wise, that offer many good products. They do attract my interest.

あるので、そういうのは気になるかな。
金子　マイナーなほうが面白いものが多いと。
原田　それもあるけど、王道はそこそこ値段するし、みんな持っているから欲しくなくなっちゃう(笑)。イイのは間違いないんですけどね。あっ、70〜80年代のスキー用ダウンベストも面白いものが多くて好き。ファッションとして作っていて変なのが多いんですよ。
金子　山登り用のものは命を守るために作られるものだからベクトルが違いますもんね。原田さんが今日持ってこられたダウンベストも、相当強烈です。この派手柄のとか(※1)どこのブランドのものなんですか?
原田　わかんない。柄は誰か画家の絵だと思うんだけど、とにかくスキー用なのは間違いないです。スキー用のダウンウェアって大体、裾らへんにリフト券のホルダーを付けるためのDリングが付いているんですよ。
金子　スタイリングも最高ですね。これはもう原田さんしか似合わない(笑)。原田さんは色と色をぶつけるから面白い。
原田　派手好きだから(笑)。内側がブルーなんで、相性のいいピンクのシャツを合わせて。上が派手な分、パンツはシックにしました。
金子　同じ派手色でも、僕のスタイリング(※2)は馴染ませる感じで組んでみました。ダウンベストを着るときは、インナーで色を拾うと合わせやすいですから。
原田　たしかにそうですね。
金子　このダウンベストもスキー用のもので、白いトライアングルトグルが印象的。なのでキャップやカットソーで白を拾っています。原田さんはエディー・バウアーのダウンベストを着たスタイリング(※3)でも色をぶつけてますね。これがまた仮面ライダーみたいな雰囲気のクセモノなダウンベストで。着こなす自信はまったくないですけど、やっぱりスタイリングが素敵だなと思いました。
原田　いやぁ、これはなかなか難しいですよね。相性のいい色を合わせてみました。

Kaneko: Does that mean you think minor brands offer more interesting things?
Harada: That's also true. But things from well-known brands are pricy and everyone has it. And that makes me no longer want it. I know they are good, though. Oh, and I like ski down vests made between the 1970 and 1980s. Many of them made as a fashion item and come in weird design.
Kaneko: Mountain climbing vests are designed to protect lives, so they have a different purpose. The vests you brought today look so eccentric. Especially this one with a bold pattern (*1). What brand is it from?
Harada: I have no idea. I guess a well-known painter drew the pattern. But I'm sure that it is for skiing, because down clothes made for skiing always have a D-ring around the hem to attach a lift pass holder.
Kaneko: I love the look, too. It only suits you. It's interesting that you always match a bold color with another bold color.
Harada: Because I love flashy things. I matched a pink shirt as it goes well with the blue color of the inner fabric. The pants are rather chic to suppress the boldness of the top.
Kaneko: My look (*2) is also in bold colors, but I made them blend in each other. When you wear down vest, it's easier if you match other colors to the color of the inner fabric of the vest.
Harada: That's true.
Kaneko: This down vest is also made for skiing and characterized by the white triangle toggles. So, I matched a white cap and white tee with it. You used lots of bold colors again in that look (*3) with an Eddie Bauer vest. The design of the item is so unique and look like a hero suit. I don't think I can style a look with it, but the way you wear it is just outstanding.
Harada: Yes, it's a bit tricky one. I matched some colors that go well with the vest.

● SPECIAL TALK with MANABU HARADA

48 DOWN VESTS

金子　原田さんにとって相性のいい色とは？
原田　僕の、というのはなくスクールカラー的なものですよね。相性のいい配色は出切ってしまっているから、あとは時代感で選ぶ。時代で色の気分も変わるじゃないですか。そういうのがあるから楽しい。色といえば金子さんのパープルのダウンベスト[※4]も、スノーライオンらしい退色の仕方がいいですよね。素材なのか、染め方なのか、絶対に褪せてる。
金子　この褪せ方にいい意味での古着の価値があると思っていて。自分、デッドストックってあんまり好きじゃないんです。
原田　僕も、未使用に古着の価値を求めていない。そんなの買ったら着られないし。だからアメリカの人がボロボロの古着に高値をつけたりするのは、イイ文化だなと思います。
金子　そう考えるとラルフローレンって凄い。新品をわざと退色させて、みたいなのを早くからやってきたわけじゃないですか。ところで原田さんのギンガムチェックのダウンベスト[※5]も個性派ですね。
原田　リバーシブルで。ウエスタンっぽい肩のヨークに時代が感じられますよね。ウールリッチの一着ですが、これも70年代か80年代の初期のものなのかな。スタイリングは、可愛くなりすぎないように色を抑えました。そのうえで、同系色の青を合わせて。
金子　内側の黄色とのコントラストもいい。
原田　金子さんのデニムのダウンベスト[※6]も70年代っぽいですよね。
金子　エディー・バウアーの一着で、色の具合がイイですよね。同じく70年代が薫るタイガーカモ柄のゆったりしたパンツを合わせて、アメリカっぽく着こなしてみました。
原田　いいですね。それにしても、この時代はなんでもデニムで作りたかったんだなぁと。
金子　まさに。じつはこの本でもリーバイスのデニムのマウンテンパーカを紹介しているんですよ(P.71)。さて、ここまで様々なダウンベストの着こなしを見せてもらいましたが、

Kaneko: What is your definition of a good combination of colors?
Harada: I don't have my own. There are some theories about it and all the compatible color combinations have been tried. So I choose colors according the mood of the time. Good color combinations vary with the times and I find it fun. Talking of colors, I like the way the purple color of your down vest from Snow Lion faded. I don't know if it's because of the material or the dyeing method, but products from this brand always faded in colors.
Kaneko: I believe the value of vintage clothing lies in the way the colors fade. I don't really like deadstock vintage clothes.
Harada: I also don't value vintage clothing that is unused. If I bought it, I cannot wear it. So I think it's great that Americans are willing to pay high prices for worn-out old clothes.
Kaneko: When you think about it, Ralph Lauren is amazing. They have been using the technique of intentionally fading new products for a long time. By the way, the gingham vest (*5) you wear in the photo looks so unique.
Harada: It's reversible. The cowboy-like shoulder yokes convey the mood of the era. It's from Woolrich and I wonder they made it in the early era, the 1970s or 1980s. I used moderate colors for the look not to make it look too cute and matched a hoodie in similar blue.
Kaneko: It creates a nice contrast with the yellow fabric inside.
Harada: Your denim down vest (*6) also has a very 70s design.
Kaneko: This one is from Eddie Bauer and I like the condition of the color. I matched baggy pants in a tiger camo pattern with a 1970s vibe for an extra American feel.
Harada: That's cool. People in those days seemed to want to make everything in denim.
Kaneko: Indeed. This book actually features a denim mountain parka from Levi's (see page 71). Well, you have

80

原田さんが思うスタイリングにおけるダウンベストの魅力とはなんでしょうか？

原田 やはりインナーにもアウターにも使えるところではないでしょうか。着丈や襟のありなしによっても雰囲気が変わるし、いろいろな着方ができる面白いアイテムですよね。

金子 おしゃれですよね。僕は今日原田さんが着ている短丈のもいいなと思いました。

原田 短丈だとタックインでベルトを見せるのもアリですよね。何にせよ何十年前と同じ定番のスタイリングではダメだと思っていて。どこかにアップデートは加えたいですね。

金子 今日のスタイリングでいうと、コーデュロイパンツの中でも、スラックスタイプの上品なものを合わせたりとか？

原田 そうですね。でも、もっと飛ばしてもいいと思います。ギンガムチェックのダウンベストを着たスタイリング(※5)でいうと、ビームス プラスのシャカシャカの白パンが飛ばした感じがあって、定番っぽくなかった。

金子 あのパンツ、じつはセットアップの組上があって。ファイヤーマンデパートメントのジャケットが元ネタなんです。原田さんはこなし方が独特ですね。

原田 新しいものはどんどん取り入れたい。

金子 チャレンジを続けていきたいということですよね。ちなみに、シャツのボタンを大きく開けていた(※1)のを見て、いわゆる"ラルフローレン開け"を思い出しました。

原田 若い子に聞くとボタンダウンシャツっておじさんっぽく映るらしいんで、そう見せないように着方を変えた感じかな。何か工夫しないと、昔のままだと思われちゃう。若い子の格好は、アップデートのヒントになります。古着屋で働いている子たちが、8年前くらいにブーツカットのデニムを穿き出したじゃないですか。で、それイイねって言い出したら、その3年後くらいにわっと流行って。

金子 モヘヤニットもそうでした。若い方にももっと、うちの服を着てほしいなぁ。

shown us a variety of looks with a down vest. What do you like about a down vest in terms of styling clothes?
Harada: I like it because it can be worn both as an inner layer and outerwear. And the look changes depending on the length and whether or not it has a collar. It's also fun that you can wear it in many different ways.
Kaneko: It looks stylish, doesn't it? I like the short one you wear today.
Harada: With a short-length vest, you can show the belt with the shirt hem tucked in. Anyway, I feel that staple looks that have remained unchanged for decades are no good. We need to update them to some extent.
Kaneko: To take your today's outfit as an example, you match a pair of slacks-type cord pants. Is that what you mean?
Harada: Yes. But I think it can be more eccentric. As for the look with the gingham down vest, white nylon pants from BEAMS PLUS give an eccentric feel to the outfit and make it look not ordinary.
Kaneko: Those pants have a matching jacket. We designed it by using a jacket for firefighters as a reference. You wore it in a unique way, though.
Harada: I'd love to incorporate new things.
Kaneko: It means you want to keep trying, right? When I saw you wearing a shirt with many buttons undone, it reminded me of Ralph Lauren's looks.
Harada: Young people say they think button-down shirts are for old men. So I changed the way I wear it to try not to look like that. If I don't put in some effort, people will think what I do is the same old thing. The way young people dress gives me a hint for updating. Kids working at vintage clothing shops began to wear bootcut denim pants about 8 years ago. I said it was cool and then the style became a trend about 3 years later.
Kaneko: The same thing happened for mohair sweaters. I want more young people to wear BEAMS PLUS's clothes.

CAGOULES

これさえあれば傘いらず
アウトドアらしい究極のレインギア

カグール。フランス圏ではさまざまなスペルで綴られるこの単語は、フランス語から生まれたイギリス英語。軽量かつ耐候性を兼ね備えた、フードが付属するレインコート、またはアノラックを指す言葉です。小雨でも傘を差す文化の日本ではあまり馴染みのないものかもしれませんが、アウトドア好きにはたまらないアイテム。ときに激しい雨風に襲われる山岳地でのハイキングに用いられるギアとあって、雨具として非常に頼もしい存在です。だから、もっと普及してほしい！ちょっぴり着づらいのはご愛敬ということで(笑)。僕が着ているのは、シエラウエストの一着。ロング丈ですが裾広がりで、動きやすいように作られています。ゆったりしたシルエットに合わせて、足元はキャンプモックでリラックス感を添えました。

No need for an umbrella.
The ultimate outdoorsy rain gear.

The British English word cagoule is borrowed from French, meaning a lightweight weatherproof hooded raincoat or anorak. In French, there are multiple ways to spell it. This rain garment is not so common in Japan because Japanese people tend to use umbrellas even in just light rain. However, outdoor enthusiasts do love cagoules. Initially invented as a garment to protect the wearer from an occasional storm during hiking in a mountainous area, a cagoule is ideal and reliable rain gear. I want more people to know how great it is! You may find it a bit tricky to put on, but nothing is perfect, you know. In the photo, I'm wearing a cagoule made by Sierra West. The A-line hem gives more freedom of movement even with long length. The camp moccasins add an extra relaxing feel to the look.

● CAGOULES

THE NORTH FACE 49 CAGOULES

70年代に作られたハイキング用カグールです。上半身は二重構造になっていて、雨だけでなく風も凌げるように考えられています。フロントに大きく付いたポケットデザインも、インパクトがあってカッコいい。裾の内側にスナップボタンが付いていて、ショートレングスにして着用することもできます。シーンに合わせて変形できる仕様というのも、ディテール好きにはたまらないですね。

This is a hiking cagoule made in the 1970s. The upper body fabric is double-layered to keep out rain as well as wind. The huge pocket on the center looks cool and gives a great impact to the design. And the hem can be shortened with press studs. Like me, people who yearn to learn the details of clothes may love such transformable design.

Sierra West — 50 STORM KING CAGOULE

シエラウエストは、カリフォルニア州サンタバーバラで1970年に創業したアウトドアブランド。こちらは足の可動域を考慮し、裾に向かってフレアさせた、イカみたいな!?シルエットが目を惹くカグール。GORE-TEX®生地を使用し、後ろ身頃は贅沢な一枚生地になっています。こちらも左頁と同様、ショート丈になる優れモノ。

Sierra West is an outdoor wear brand established in Santa Barbara, California, in 1970. This cagoule comes with a characteristic long, A-lined silhouette to allow for greater freedom of leg movement. It features GORE-TEX® and the back panel is extravagantly made of a piece of cloth. The length can also be shortened by folding and buttoning the hem, just like the one on the left page.

● CAGOULES

Alpine Designs 51 NYLON CAGOULE

現ブランド名に改名する前、アルプ スポーツ時代のクラシックムード満点なカグールです。60年代に作られた一着とあって、用尺使いが贅沢。肩周りが1枚生地でシームレスになっているため、着心地が抜群にいいんです。ドレープの表情も素晴らしいですね。ブランドらしいカラーも気に入ってます。

This cagoule filled with classic vibes is from the era, the brand's name was Alp Sport. The extravagant use of fabric speaks to its creation in the 1960s, evoking a classic mood. The seamless shoulder panel is made of a single cloth and feel extremely comfortable. Along with the beautiful drapes, I like this color that symbolizes the brand.

Gerry 52 GERRY PARKA

1947年に発売され、K2やエベレストの登山者にも着用されたパーカです。1964年には、商品名が"MOUNTAIN PARKA"に変わりました。"マウンテンパーカ"という呼び名の人気が窺えます。素材は、ハイスペックなデュポンのナイロンを用いた軽量防風クロス。縦長のシルエットに漂うクラシックムードがたまりません。

This parka was first produced in 1947 and had worn for K2 and Everest expeditions. It changed its name to Mountain Parka in 1964. I guess the word "mountain parka" itself gained popularity around that time. The fabric is lightweight windproof cloth made of Du Pont's nylon. The vertically elongated silhouette gives an extra classic mood.

Blacks 53 ANORAK VENTILE® MODEL

イギリス名門ブラックスの一着は、60年代の代表的なデザイン。イギリス軍のカグールをベースにしたことが窺えます。当時2つのモデルが展開されていましたが、こちらは生地にベンタイルを用いたハイスペックなモデル。ベンタイルらしいハリのある風合いと長年着込んだことで生まれた味は唯一無二です。

Blacks is an established British brand. This anorak comes in typical 1960s style and was probably designed based on the British Army's cagoule. The brand offered two different models and the one in the photo is the top-tier VENTILE® model. The VENTLE® fabric feels crisp and has a one-and-only worn-out appearance.

Recreational Equipment, Inc. 54 CO-OP ANORAK

イギリスのアノラックをベースに製作したであろうモデル。60年代に存在した希少な日本製の一着です。サイズは特大のLL。生地はコットンで柔らかく、防水性はありません。あくまでカグールをモチーフにした服ですが、ジップの配色を赤にするあたりにREIのアメリカンな個性を感じて、愛おしいですね。

This model is assumed to be designed based on the British anorak. This one here is a rare piece from the 1960s and made in Japan. The size is written as LL, it's actually huge. The cotton fabric feels soft and is not waterproof. It is just a piece of clothes that is designed like a cagoule. The red zipper reflects REI's unique American individuality.

FIELD PARKAS

背景に思いを馳せる
個性派揃いのフィールドパーカ

ダウンなどのインサレーションのないフィールドパーカも、アウト
ドアカルチャーを語る上で欠かせない存在。ダウンウェアよりもデ
ザイン自由度が高く、多くのブランドが魅力的なそれをリリースし
てきました。着心地を重視した結果なのか、生産性を高めるためな
のか、60年代の古いものには生地の用尺を無視して肩周りを贅沢
な一枚取りにしたウェアが見られ、これまた背景を想像するのが楽
しい。80年代にはアウトドア用なのに白!?なんてモデルも登場。
ギアからファッションアイテムへ変遷した歴史を辿れるあたりも、
とても興味深いです。写真の一着は、サバイバロンのセーリング用
パーカの名品。手旗信号フラッグ柄のシャツやパンツを合わせ、気
分を盛り上げました。足元は当然、トップサイダー一択です。

Unique field parkas
that stir imagination

A field parka is also an essential item for outdoor culture. Because it is more flexible
in design than insulated garments, many different brands have offered fascinating
field parkas. I'm not sure whether it is for wearability or productivity, but there are
some old 1960s parkas with a shoulder panel extravagantly made of a single piece of
cloth. It is fun to think of why they did so. In the 1980s, a brand made a white field
parka. It might be designed not for outdoor use but as a fashion piece. It's also
interesting to trace such a change of the purpose. The one I'm wearing in the photo
is Survivalon's famous sailing parka. The flag pattern shirt and pants give an extra
sailing mood. Top-Sider shoes are the only choice for this look.

●FIELD PARKAS

THE NORTH FACE made by SIERRA DESIGNS 55 STANDARD MOUNTAIN PARKA

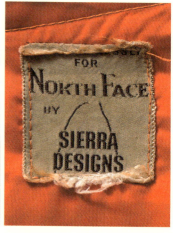

1966年創業のザ・ノース・フェイスのモデル。当時シエラデザインズが生産を手掛けたと思われる一着。ネームタグも併記されています。生地は60/40クロスではなく、高密度のナイロン。メタルジップとビスロンファスナーを併用していたり、アウトドアウェアの過渡期を感じさせてくれる、希少で貴重な一着ですね。

The North Face, a renowned outdoor wear brand established in 1966, designed this parka. There is, however, Sierra Designs' name on the tag, because the company seemed to manufacture the brand's products back then. This one is not made of 60/40 cloth, so the usual nylon fabric was used for this piece. The combined use of metallic and plastic zippers reminiscent of the transition period of outdoor wear. It is truly a rare and valuable piece.

SIERRA DESIGNS 56 MOUNTAIN PARKA REEVAIR MODEL

レザーのガンパッチがカスタムされ、映画『ディア・ハンター』の世界観を想像させるブランド初期のマウンテンパーカ。生地はナイロンで、古いモデルの象徴である極小のスナップボタンにはまだブランド刻印が見られません。ポケットフラップの形状やライニングカラーも、現在のモデルとは異なっています。クラシックな雰囲気がたまらない一着ですね。

Sierra Designs made this mountain parka in the early of their brand history. The leather gun patch reminds me of the movie "Dear Hunter". Made of nylon fabric, this parka features tiny press studs without the logo, details that suggest it's an older model. The shape of the pocket flaps and the color of the lining are different from the latest model, and the classical feel of this item does fascinate me.

KELTY 57 MACKINTOSH ALL WEATHER PARKA

バックパックの雄であるケルティも、実はこだわりのマウンテンパーカを作っていました。こちらは、GORE-TEX®素材を用いたモデル。ステッチ部分の樹脂は、シームテープが誕生する前という証です。ヨークや、複雑なポケットなど随所に手が込んでいて、アクが強い(笑)。モデル名も"マッキントッシュ"と張り切っていて、そのアピール感も愛おしい。

Kelty is known for their backpacks, but they used to offer elaborate mountain parkas, too. This GORE-TEX® model features resined seams, and that means it was made before seam tapes were invented. It is intricately crafted with a yoke and detailed pockets, and has a too strong character. As the brand named it Mackintosh, they must be so confident about it. I like that, though.

● FIELD PARKAS

Gerry 58 MOUNTAIN PARKA

肩周りにかけて縫製がないシームレスな生地使いの一着。このパターンは古いモデルの特徴で、表地と裏地が共布というのも、輪を掛けて贅沢な仕様。この時代のジェリーは個性的なディテールが多く、いつも魅力を感じますね。同じモデルでターコイズもコレクションしています。

This mountain parka is designed without seams around the shoulders. It is one of the characteristic details of old mountain parkas. The lining is extravagantly made of the same fabric as the shell. Gerry offered many products featuring unique detailing around that time and that always fascinates me. I own another one in turquoise color.

Recreational Equipment, Inc. 59 MOUNTAINEER STORM PARKA

こちらもシームレスで贅沢な生地使いが特徴的な、70年代のモデル。着心地はもちろんのこと、シームレスな縫製仕様によって生み出される着用時のドレープが最高なんです。この時代の王道とも言えるポケットデザインや、ツバを設けたフードのディテール、表裏の配色もお気に入りの一着です。

This storm parka was made in the 1970s and is also characterized by the seamless design and extravagant use of fabric. Not to mention the comfort, the seamless body creates beautiful drapes. The pocket design often seen on the products made in the period, brimmed hood and color combination are all my favorite, too.

Early Winters 60 TERRASHELL

GORE-TEX®といえば、のアーリーウィンターズより、配色が最高な一着。ハの字の大容量ポケットは、カンガルーポケットからイメージされています。カタログには"マウンテンパーカとアノラックのいいとこどりをしたようなデザイン"と記載も。独自のダブルドローストリングシステムのフードディテールは秀逸です。

When it comes to GORE-TEX®, you must not forget Early Winters. The color combination of this jacket is just outstanding. The huge inverted V-shaped pocket is designed with inspiration from kangaroo's pocket. The catalog says the design has the best of both mountain parkas and anoraks. The brand's unique double drawstring system is featured on the hood.

WOOLRICH 61 MOUNTAIN PARKA

80年代に作られた白のマウンテンパーカ。山で着る服なのに視認性が低く汚れやすい白というのは、これがファッションアイテムとしての立ち位置だった現れといえるでしょう。ともあれこのバーガンディーとの組み合わせというクリーンなルックスには、東海岸ブランドらしい洗練さを感じます。

This white mountain parka was produced in the 1980s by the East Coast brand, Woolrich. The color is not ideal for mountain activities because it has low visibility and gets dirty easily. That means it was designed as a fashion item. However, the color looks neat and clean with burgundy, conveying the sophistication of the East Coast.

●FIELD PARKAS

L.L.Bean 62 SUMMIT PARKA

クラシックな防水素材、ベンタイルを用いてヨーロッパのフィールドジャケット型に仕上げた希少なモデルです。アシンメトリーな胸のポケットや収納できるフードなど、ディテールにも魅力満載。アメリカ製とイギリス製が存在しますが、本品は後者。イギリスらしい佇まいが魅力的な一着です。

VENTILE®, classical waterproof fabric, is used for this summit parka to make it like a European field jacket. This rare piece comes with many fascinating details, including asymmetric chest pockets and a packable hood. This model was manufactured both in the U.S. and U.K., and this one is the latter. I love the British atmosphere it has.

Alpine Hut 63 SKI PARKA

50年代のヴィンテージ。メタルファスナーにメタルストッパー、肩周りの贅沢な生地取りが特徴的で、刺繍タグにもクラシックなムードが漂います。ステッチ裏には雨水の浸入を防ぐための樹脂のコーティングが。シームテープ誕生以前の技術が見られる点にも、価値があります。

Vintage piece from the 1950s. The design is characterized by the metallic zipper and stopper, extravagantly used fabric around the shoulders as well as the classical embroidered tag. The seams are resined on the back side to prevent rain and water from entering. It is the technique used before seam tape was invented and make this item historically valuable.

Survivalon 64 SAILING JACKET

80年代のセーリングジャケットで、マイティマックにマクレガー、アバクロンビー&フィッチと、名だたるブランドが同じデザインのそれを作っています。フードが収納できる仕様やサングラスポケット、リバーシブルなど、ブランドならではの機能美とデザインの妙がお気に入りです。

Survivalon made this sailing jacket in the 1980s, and many renowned brands such as Mighty-Mac, McGregor and Abercrombie & Fitch also offered their own jacket in the same design. From the packable hood to shade pocket to reversible design, I like the functional beauty and ingenious pattern that the brand could only create.

Rocky Mountain Featherbed 65 MOUNTAIN JACKET

ネイティブアメリカンのレザーケープを元にしたヨークに、ココらしさが光るジャケット。使用されている生地はそれぞれダック(左)とGORE-TEX®(右)。後者はネックにフードが内蔵されています。レザーとGORE-TEX®というアメリカらしい組み合わせがお気に入りで、雨の侵入を防ぐためか、胸ポケットの方向が違うのも面白いですね。

The yoke inspired from Native American's leather cape characterizes these jackets and the brand itself. One on the left is made of duck fabric, while GORE-TEX® is used for the one on the right. A hood is packed in the collar of the latter. I like the way they match leather with GORE-TEX®, because it looks so American. It is also interesting that the chest pockets are placed in an unusual way probably to prevent rain from coming in.

THE OTHER
MY COLLECTIONS

掘れば掘るほど沼にハマる
アウトドアギアの世界はこうも奥深い

ここからは、僕のお気に入りのアウトドアギアコレクションをジャンルを跨いで紹介します。アウトドアギアの面白さは、それぞれに用途があり、目的をなすためのディテールや機能美を備えているところ。異なるブランドから似たアイテムがリリースされていたり、工場は同じなのにディテールがちょっぴり異なるアイテムがあったり、深堀りするうちに興味深い事実に気づくこともあります。アメリカは常にアップデートをする国であり、1つのアイテムを年代別に見比べることでその変遷が見える面白さも。そんなこんなで、コレクションは増え続けるばかりです。写真で羽織っているのは、カモをおびき寄せるおとり＝デコイを持ち歩くための服。獲物からバレないよう!?全身をハンティングパターンでスタイリングしてみました。

The more you learn, the more you want to know.
The depthful world of outdoor gear.

In this section, I will introduce the collection of my favorite outdoor garments. I find outdoor gear interesting because every single item has a purpose and is designed with details and functions for the purpose. If you further your knowledge, you may detect some curious facts such as different brands made similar products and garments manufactured in the same factory come with slightly different detailing. Because America loves improvement, you may also find out differences when comparing a product by the year of manufacture. With this thing and that my collection keeps expanding. What I wear in the photo is a garment designed to carry decoys. I'm in the all-camo outfit so that hunting game cannot find me!?

● THE OTHER MY COLLECTIONS

66 LONG BILL CAPS

ハンティングやフィッシングの定番で、かぶるだけでアウトドアの趣を演出できるロングビルキャップは、僕の大好物。ギアというよりファッション感が強い柄やネオンカラーのものもたくさん持っています。ハマったきっかけは、カモのくちばしのような形状のツバを備えたそれ(写真中央下あたり)で、獲物のカモに似せるためにくちばしの形状をしているのだそう。NYで見つけて、これは面白い!!と感動し、気づけば膨大な量のコレクションになっていました。そんな自分の傾倒した趣味からアメリカの老舗クエーカー マリン サプライと繋がり、今ではビームス プラスで共にMade in U.S.A.の物作りができていることは冥利に尽きますね。

Long bill caps are classics for hunting and fishing, creating an outdoorsy mood just by wearing it. I love the items and own so many of them, including more fashion-centered ones in bold design and neon colors. That one with a brim like a duck bill (a cap shown in about the center of the photo) got me hooked on long bill caps. I was told that the brim is designed to imitate the winged game. I found it in New York and so impressed that I have collected quite many of them. My personal preference later led me to Quaker Marine Supply, and I luckily could have an opportunity to work with them to make U.S. made caps for BEAMS PLUS.

● THE OTHER MY COLLECTIONS

67 FIELD HATS

フィールドハットは、とにかく道具としての機能美を兼ね備えていて面白い。たとえばベンチレーション機能の菊穴から雨が入らないようにカバーが付いているサファリハットや、通気性を高めるべくベンチレーションの穴が8つも付いているフィッシングハットなど。ベージュやオリーブなどアースカラーがほとんどですが、それぞれに個性があるから、つい集めてしまうんですよね。

I find field hats interesting because of the functional beauty. For example, a safari hat features a cloth to cover ventilation holes to prevent rain from entering and a fishing hat with 8 ventilation holes for better breathability. Most of them are in earthy colors such as beige and olive, but each item is uniquely designed. And that is why I cannot help myself collecting them.

Willis & Geiger 68 SAFARI

ウィリス アンド ガイガーは1902年に北極探検家ベン・ウィリス氏によって設立されたブランド。探検隊や米軍コントラクターのほか、様々なブランドのOEM生産も手掛けてきました。サファリクロージングも得意とするところ。ポケットワークしかり、ワイルドな風貌とは裏腹の東海岸ブランドらしい上品なライニングしかり。大げさなディテールはネタの宝庫です。

Willis & Geiger was established in 1902 by Arctic explorer Ben Willis. They have outfitted expedition parties and U.S. military forces, while manufacturing OEM products for various brands. Their safari clothing is known for its quality. From the pocket layouts to chic lining with the East Coast vibe despite the rugged appearance, their exaggerated detailing has inspired me a lot.

● THE OTHER MY COLLECTIONS

69 HUNTER YELLOW
SAFETY YELLOW

ハンティングウェアの定番カラーといえばオレンジ。ですがじつはハンターイエロー(またの名をセーフティーイエロー)というカラー名もあり、オレンジ同様、派手でありながら獲物の四足獣は視認できない誤射防止カラーとして用いられてきました。そもそもイエロー好きな僕にとってはたまらない存在。

Speaking of the color of hunting wear, you may think of orange. But hunter yellow, also known as safety yellow has also been used because it increases visibility to help prevent misfires and four-footed game animals cannot recognize the color. I personally like yellow, so these jackets do fascinate me with the coloration a lot.

●THE OTHER MY COLLECTIONS

FILSON 70 MACKINAW CRUISER

"森のワークウェア"を象徴するモデル。アメリカ北西部開拓時代から、伐採業者、木材巡視員、森林系部隊員ら森のワーカーに愛されてきました。複層的なポケットは、チョークやライト、コンパスなどを仕舞うためのもの。腰ポケットの左右の高さが違うものは機能を優先させた結果で、通称"ガチャポケ"と呼ばれる旧仕様です。柄合わせの適当さはご愛敬。

These Filson's shirts symbolize the "woodland's workwear". This model has been loved by forest workers, patrolmen and forest guards since the American frontier era. The layered pockets are designed to store various things such as chalks, a flashlight and compass. The height change of the right and left waist pockets are seen on the older models to improve usability. Poor pattern matching? It looks rather charming for me.

Hercules 71 DUCK COATS

1961年に発行されたシアーズのカタログに掲載されている、ブランケット付きのチョアジャケット。レザーパイピングが目を惹く一着は、ハンティング用にカスタムされたもの。背中には巨大なパッチ跡が有り。当時のオーナーのカスタムセンスには脱帽です。なおヘラクレスは、シアーズのワークウェアブランド。ハンティングとワークのミックス感がたまらない。

According to a Sears's catalog published in 1961, this chore jacket came with a blanket. This one here is customized for hunting and characterized by the leather piping. There is a huge patch mark on the back. The aesthetics of the previous owner are just mind-blowing. Hercules was a workwear brand run by Sears and they mixed the elements of hunting and workwear in a very fascinating way.

● THE OTHER MY COLLECTIONS

L.L.Bean 72 OLD L.L.Bean

エルエルビーンの原点であるハンティングやフィッシングなどのアイデンティティを、色濃く表すOLD L.L.Bean。どれも20年代から30年代のヴィンテージアイテムです。クラシックなウール地×格子柄、カモ柄のブランケット、レザーキャップ、背中にフィッシュバッグを取り付けるディテールなど、悠久の時を感じさせるコレクションですね。

Old L.L.Bean clothes vividly reflect the brand's identity of hunting and fishing. All the items on this page are vintage pieces made between the 1920s and 1930s. From the classical plaid woolen coat, blanket with a duck pattern, leather cap and jacket with an attachment for a fishing bag, every products here make me feel the passage of eternal time.

L.L.Bean 73 RAINBOW LAKE JACKET & PANTS

映画『ラスベガスをやっつけろ』でラウル役のジョニー・デップが着ていたジャケットと同柄のアイテム。大胆なパッチワークに、70年代の空気感が表れています。当時エルエルビーンだけではなく、ウィリス アンド ガイガーとアバクロンビー&フィッチにも同様の製品が見られました。同時期に複数のブランドから出ていたということは、製造元が一緒なのか!?　面白いアイテムですね。

This jacket has the same pattern which Johnny Depp as Raoul Duke wears it in the film "The Fear and Loathing in Las Vegas". The suit was designed by L.L.Bean with bold patchwork detail that reflect the mood of the 1970s. Willis & Geiger and Abercrombie & Fitch offered the similar products around the same time, and that probably means they used the same manufacturer…? Truly intriguing.

● THE OTHER MY COLLECTIONS

Red Head 74 LINER VESTS

カモやキジの絵柄が描かれた、大好物のライナーベストです。フィッシングやハンティング用品を扱うブランドのもので、柄使いが抜群にイイ。ライナーベストでこの手の柄というのも珍しいですよね。おそらくブランドがオリジナルで製作した柄だと思うのですが、そこがまた所有欲をくすぐる。この2着は私の中の柄大賞。

These liner vests are adorned with a game bird pattern and strike a chord with my aesthetics. Red Head is a brand offering outfits for fishing and hunting and their use of patterns are just amazing. It's not common to see such patterns on liner vests. I think they are the brand's original patterns and do stimulate my desire of possessing. These are some of the best patterns for me.

75 POLISH ARMY CHEETAH CAMOUFLAGE

あるとは聞いていたけれど実物を見たことがない。そんな伝説の通称"チーターカモ"に、とうとう出会えました！ 60年代のポーランド軍のファティーグジャケット。チーターの模様とカラーリングが相まって、森の中でしっかりカモフラージュしそうですよね。タイガーやリザードなどアニマル柄のカモフラージュはいろいろありますが、知られざるカモが世界にはまだ存在するのかと思うと、ワクワクしてしまいます。

I had heard about there was a camo pattern called "cheetah camo" but never seen it. And finally, I could own a fatigue jacket adopted by the Polish military in the 1960s. The cheetah pattern and the color combination seem to camouflage you well in forest. From tiger to lizard, various animal-inspired camo patterns have been created, so there must still be unknown ones in the world. That sparks my curiosity.

● THE OTHER MY COLLECTIONS

76 LINER JACKETS

左からイデアル、ノーブランド、レッドヘッドの製品。ライナージャケットはコートの下に着る防寒着ですが、個性的なものが多く、ついつい欲しくなってしまうんですよ。左の一着はジグザグのキルトパターンが面白く、中央の一着は鳥とハンター柄のムードが良い。右の一着はハニカムの織り柄がカッコよくて、どれも服作りで参考になるアイテム達です。

These liner jackets are made by, from the left to right, Ideal, an unknown brand and Red Head. Liner jackets are designed to wear underneath a coat to prevent cold, but many of them are so uniquely made and I can't help myself getting them. The one on the left comes with an unusual zigzag pattern, while the bird and hunter pattern on the one in the center has a good quaint mood. The honeycomb pattern of the one on the right looks cool, too. All of these sparks my creativity for making clothes.

Bradley Mountain Wear 77 FLEECE JACKETS

サンタフェのローカルブランド、ブラッドリー マウンテン ウェアの製品です。これが、じつに個性的。真骨頂はジップを3つ使ったベンチレーションで、腕から脇下を通って身頃まで開口できるようになっています。また生地の切り返しや配色も最高の一言。小規模生産だからこそのこだわりを、随所に感じます。

These jackets are made by Santa Fe-based brand Bradley Mountain Wear. The design is truly one of a kind. They use three zippers for the ventilation system that stretches from the arm to the waist. The patchwork design and color combination are both amazing, too. There are many elaborate details that could only be realized in small lot production.

● THE OTHER MY COLLECTIONS

Early Winters 78 GORE TRAINER SUITS

GORE-TEX®のパイオニアはこんなモノまで作っていたの?と思わず驚いたトレーナースーツ。ジャージー生地はもちろん、GORE-TEX®仕様。雨から頭も守るべく、フィールドパーカ風のアンバランスなフードが付属しています。それにしても、ランニングウェアなのにハイスペックすぎますよね。でもそんなところも嫌いじゃないんです。

When first seeing this trainer suit, I was surprised that Early Winters, the pioneer of GORE-TEX®, has made even a product like this. The jersey fabric is made of GORE-TEX®, of course, and the item comes with a field parka-like mismatched hood to protect the head from rain. Isn't it too high-tech for running wear? I like it, though.

L.L.Bean 79 REFLECTOR CYCLING JACKET

リフレクター素材を使った80年代のサイクリングウェア。当時はランニングウェアにも同素材が使われていました。撮影時にフラッシュを焚いて気づいたのですが、腕だけリフレクターが剥げている。どう着ればこうなるのか？想像するのも面白いですね。また、あえてのバティック柄というのもアガリます。当時のアーリーウィンタースにも同生地を使ったサイクリングウェアが存在していました。

These cycling jackets and caps from the 1980s are made of reflective fabric that used to be chosen for running outfits back then. When taking the photos with a flash on, we realized that the reflectors around the arms are came off. But how did it happen? It is fun to think over the reasons. And the use of a batik pattern makes me excited. Early Winters offered cycling clothes made of the same fabric around that time.

● THE OTHER MY COLLECTIONS

80 SKI

ブランド不詳のシャツは、おそらく50〜60年代に作られたもの。ホリデーカラーなのか黒×赤のコットンジャカードで表現されたスキー柄がカッコよく、柄作りの参考となる一着。奇妙な形のニット帽は"イカ帽"と呼ばれるそれで、地域色が出るところが面白い。ウィンタースポーツムードの演出に重宝しています。

This skiing jacket was probably made between 1950s and 1960s by an unknown brand. The holiday-ish black and red jacquard pattern looks so stylish and inspires me to make clothes. It is interesting that the knitted hats convey the mood of a place it was made. They are ideal accessory to add a sporty vibe to your look.

Mary Maxim 81 COWICHAN SWEATERS

カナダはマニトバ州シフトンにて30年代に創業した、メアリー・マキシムのカウチンセーター。50年代のアメリカにおいてキット通販がヒットしたメーカーなので、これらはおそらく購入した誰かが手で編んだもの。ジグザグした柄の味もさることながら、作る人によるヘタウマ感が面白い。当時のアメリカを感じさせる柄とその豊富さが魅力的ですが、中でも僕は生き物柄が好きかな。

These Cowichan sweaters were designed by Mary Maxim, a company established in the 1930s in Sifton, Manitoba, Canada. They made big hits with their mail-order kits in the US during the 1950s, so these sweaters may be handknitted by their customers. Along with the chunky stitches, each pattern looks crude but charming. All of them convey the mood of the era and are so fascinating. My favorite patterns are ones with living things.

● SPECIAL TALK with SETSUMASA KOBAYASHI

82 GEAR VESTS

I GOT BLISTERS
ON MY FINGERS

Gear vests are fantasy!

小林節正
Setsumasa Kobayashi
Mountain Research / Founder and Designer

1961年生まれ。1993年にシューズブランドのセット、フ
ァッションブランドのジェネラルリサーチをスタート。
2006年に『……リサーチ』プロジェクトを始動する。ア
ウトドアブランドのマウンテンリサーチはその中核だ。

Born in 1961. After launching his own shoe brand, Sett and
apparel brand, General Research in 1993. He started his
"……RESEARCH" project in 2006. His outdoor wear brand
Mountain Research is the core of the project.

116

ギアベストとは、ロマンであり"ファンタジー"

金子　小林さんの事務所はいつ来ても凄い！フィッシングベストだけでこんな量ですもん。

小林　同じものでS、M、Lあったりするからさ。各サイズないとフィットがわかんないでしょ？　釣りで着るわけでもないのにねぇ。

金子　小林さんのコレクション見ると、自分は大したことないなぁと思います。

小林　いやいや、金子は古着がいい値段になってから集めてるから、そこは俺よりも気合い入ってるよね。俺の場合、二束三文みたいな値段で売ってる時代に集めたやつだから。

金子　いえいえいえ、相当気合い入ってないとこんなに集まらないですよね。

小林　その昔、四畳半の部屋にコレ全部を仕舞っていたのは信じられないね、確かに(笑)。

金子　古着は、いつ頃集め出したんですか？

小林　それ聞かれると困っちゃうんだけど、どうしてこういう状態になってるかっていうと、自分は元々靴屋だったんですね。で、服のデザインを1994年あたりに始めるんだけど、右も左も洋服の作り方がわからないのよ。ベルトをこうつけて、っていうデザインアイデアはあるんだけど、いざフィットって言われると、え？って。そんなわけで、パタンナーに説明するときに具体的なもの…つまり元ネタの実物やサイズサンプルを持っていれば、どうにか対処できるのでは？と考えたわけ。

金子　なるほど。

小林　デザイナーたちはきっと何かをリサーチして服を作っているのだろうし、ならば自分はデザインのリソースたるドンズバの元ネタを手に入れて、それを見せて説明する、というやり方でやってみようかと。だから古着が洋服作りの先生代わりだった。説明するのにバラさなきゃいけないこともあるから、同じものを複数買ったりしてね。気づけばどんどん増えていたと、こういうわけ。仲間に譲ればすっきりするけど、レコードと一緒で、他人のを100枚貰っても嬉しくないでしょ？

金子　僕は嬉しいですけどね(笑)。ちなみに、

Kaneko: Your office amazes me whenever I come! You have this amount of clothes only for fishing vests.

Kobayashi: Some of them are same but different in size. You can't tell the fit with just one size, you know. I don't wear them for fishing, though.

Kaneko: Your collection reminds me that I still have a long way to go.

Kobayashi: No, no. That's just because you began collecting vintage clothes after the value got hiked. You are more enthusiastic than me. I bought mine when they were so cheap.

Kaneko: No, no, no. You cannot collect so many of them without boundless enthusiasm.

Kobayashi: Actually, I can't believe I once put all of them in a tiny room which was only 8.1 square meters.

Kaneko: Around when did you start collecting vintage clothes?

Kobayashi: I don't have an exact answer for that, but the reason why I have so many clothes is that I originally made shoes. Then I began to design clothes around 1994, but I had no idea about how to make clothes. Though I did have some design ideas, I didn't know how the fit should be or something like that. So I came up with an idea that I might be able to deal with it if I show a piece of actual clothes to explain a pattern maker what I want to make.

Kaneko: I see.

Kobayashi: I guessed that designers design their clothes by researching something. Then I thought that I should get my inspiration source, an actual vintage garment to explain about my idea. So vintage clothes were my teacher of clothing design. I sometimes bought several of the same item when I needed to unsew one of them to show the construction. And the number of clothes I own had been increased before I knew it. If I give them to my friends, this room feels refreshing. But, just like records, nobody would be happy if he received 100 pieces of clothes from others.

Kaneko: I will be happy though. Do you remember the first vintage item you bought?

SPECIAL TALK with SETSUMASA KOBAYASHI

82 GEAR VESTS

一番最初に買ったものって覚えてますか？

小林 さすがに覚えてないよー。でも初めて買った釣りベストはコロンビアのこれ(※1)。当時は池袋西武にスポーツ館があって買ったの。今見ても、くどいくらいのポケット！

金子 へぇ、これだったんですか。

小林 ポケットが便利だったから、当時はよく海外に着いってたよ、ヨーゼフ・ボイスの真似ってものあるけど。ところがさ「お前も釣りやるのか？」なんて釣り人のおじさんに話しかけられたりして、面倒だったね(笑)。

金子 当時フィッシングベストをファッションで着る人、いなかったんじゃないですか？

小林 そうだよね。1982年とか83年とかだから。こんだけポケットついてるってすごいでしょ？　売り場でめちゃくちゃ熱くなっちゃって。日本じゃ他では売ってなかったの。

金子 この一着から収集癖に火が付いて、止まらなくなっちゃったんですね。

小林 そうそう。だってほら、この時代は細かい仕様の違いでバリエーションがいろいろ出ているから。80年代の後半あたりになるとポケットの縁が丸くなったりして。

金子 気持ちわかります。ベストはサックス(※2)のも結構お持ちなんですね。

小林 サックスのはカメラマンベストだね。後ろに三脚が挿せる用になっていて、毛鉤を付けるムートンのフサフサがない。あとは、フィッシング用でもファッションマーチャンダイズ化されたカラフルなの(※3)があって、それはそれで好き。

金子 これ、同じのを全色揃えたんですか？

小林 いや、多分あともう2、3種類あったと思う。買いそびれたのもあるし。

金子 これらをご自身の服作りに活かしているということですよね？

小林 そうだね！　洋服屋として1994年から仕事を始めたんだけど、1997年の秋冬になるとパラサイトというテーマを掲げて、海で見るフジツボみたいに、たくさんのポケ

Kobayashi: No, that's impossible. But the first fishing vest I bought was this (*1). The Ikebukuro Seibu department store had an annex for sports-related products back then, and I got it there. I thought it had too many pockets and still think the same!

Kaneko: Oh this was the first vest you bought.

Kobayashi: I often wore it abroad around that time. I wanted to look like Joseph Beuys, but I often got asked by unknown anglers, "Do you do fishing, too?", which was a bit annoying.

Kaneko: I guess no one wore a fishing vest as a fashion statement at that time.

Kobayashi: Yeah, it was in 1982 or 1983. Don't you think it's amazing that a vest has so many pockets? I was so excited in the department store. Only there you could get it in Japan at that time.

Kaneko: This one ignited your collector's spirit and the passion hasn't stopped since.

Kobayashi: Yeah. That's because there were a few different versions for one item around that time. The design slightly differed. For example, the edge of the pocket was rounded in the end of 1980s.

Kaneko: I understand your feelings. You have many vests in saxe blue (*2).

Kobayashi: This one in saxe blue is a photographer vest. It has an attachment for a tripod on the back and no wool fly patch. Some fishing vests are designed in vivid colors (*3) for the general market. I do like them as well.

Kaneko: You bought all colors available?

Kobayashi: No, there were a few more colors that I couldn't get.

Kaneko: So you incorporate these into your own clothing design?

Kobayashi: Yes! I started my clothing design business in 1994 and, for my 1997 fall/winter collection, I created clothes with a lot of barnacle-like pockets on the theme of Parasite. My inspiration at the time was fishing vests.

トに寄生された服(※4)をシリーズで作ったんだけど、発想はフィッシングベストだったよ。

金子　やっぱりすごいなぁこれ、とんでもない服です。

小林　縫製工場にとことん嫌われるよね(苦笑)。縁とお付き合いに救われたわけだけど。

金子　フィッシングベストといえば、自分はやっぱり、ポケットとかディテールの面白さに惹かれます。

小林　うんうん。そして機能ではあるんだけど、つまるところレイアウトの妙なんだよね。グラフィカルな美しさ。自分は釣りをしないから毛鉤の箱の大きさとか知らない。だから、このポケットにはどんなモノを入れるのだろうか？と心を巡らせるのも楽しいし、ファンタジーとして楽しんでいる部分もあるかな。

金子　ファンタジー！　わかります。ファンタジーの世界からリアルな釣りを楽しもうとは思わなかったんですか？

小林　いやぁ、これがさっぱりで。誘われてわざわざ北海道まで行ったりしたことあるけど、やっぱダメだった(笑)。

金子　待つのが駄目なんですか？

小林　駄目だし、釣れた魚はみんなリリースするでしょ？　その道の奥深さをひとつもわかってないからいけないんだけど、何のために釣るの!?とか、つい思っちゃうのよ(笑)。

金子　いやぁ、デザインの参考にするためだけにこれだけの量を集めたのは凄いですね。2019年にはビームス プラスとマウンテンリサーチで文字がたくさん書かれたツールベストを作りましたよね(※5)。

小林　そうそう。元ネタの１つはディッキーズのシャツでね。おそらくセールスマン用サンプルなんだけど、ボタンが2つここに付いてます、とか、マシンウォッシャブルですよ、とかセールストークになる文言が全部、シャツの上に文字で記されてあって面白いなと、長年思ってたんだ。もう1つは、50〜60年代の軍モノのサバイバルベスト。そのベストに

Kaneko: This is amazing. This is really an incredible garment.
Kobayashi: You will be completely disliked by the sewing factory. I was just lucky to have a good relationship and connection with them.
Kaneko: Speaking of fishing vests, the unique pockets and detailing always fascinate me.
Kobayashi: I agree. The reason I find it interesting is how it is laid out. It's visually beautiful. I don't do fishing and I don't know how big a fly box is. So, for me, it is fun to think of what the pockets are made for. I partly enjoy it as a fantasy.
Kaneko: A fantasy! I do understand. But have you ever thought about actually enjoying fishing?
Kobayashi: No, I'm not good at it. I was invited to go all the way to Hokkaido for fishing once, but it didn't help.
Kaneko: Is it because you can't wait?
Kobayashi: No, I can't. And all the fish you catch will be released, won't it? Because I'm totally ignorant about the cultural depth of angling, I would start thinking like, "So what are we doing it for?".
Kaneko: It's amazing that you have collected so many clothes just to use as design references. In 2019, BEAMS PLUS and Mountain Research made a tool vest on which a lot of words are printed (*5).
Kobayashi: One of our design sources was a shirt from Dickies. It's probably made as a sample for a sales person to sell the item and lots of sales pitch phrases like, "There are two buttons here" and "This is machine washable" are printed all over the shirt. I had thought it was interesting for a long time. A military survival vest from 1950s or 1960s was another design source. Each pocket of the vest has a stamped description of what survival kit is stored in it and looks so cool!

● SPECIAL TALK with SETSUMASA KOBAYASHI

82 GEAR VESTS

は、どのポケットに何のサバイバルキットが収納されているかっていうのが文字でいちいちスタンプされているんだけど、それがまた凄くかっこいい！一緒に作ったベストでは、それらをヒントにして、ミュージックプレーヤー専門のポケットには"歌"や"音"や"声"が入ってるとか、ちょっぴりひねりを加えながら言葉をスタンプしたよね。

金子　これもいわゆるファンタジーですよね。ちなみに、洋服のデザインをしていてこんなに文字を入れたことってありますか？

小林　ないない。でもいいデキでよかった。

金子　今見てもかっこいいですよね。これは時代を先取りしていたと思います。小林さんは軍用のベストもたくさんお持ちですよね？

小林　ノルマンディー上陸作戦のときに作られたベスト(※6)とか。

金子　これ映画『プライベート・ライアン』にも出てきますもんね。

小林　そうそう。いわゆるアサルトベストと言われている典型的なベストだよね。あと、ツールベストといわれるタイプのものでは、イギリス軍の機関銃のバナナマガジンを収納するこれ(※7)も好きだな。30〜40年代のものだけど、まだしつけ糸が残ってる。

金子　わぁ凄いですね。曲線も個性的。

小林　第二次世界大戦あたりのイギリス軍のモノって、どこか手工芸品を感じるんだよ。アメリカのジーンズみたいのとは真逆の。仕立て文化がまだ幅を利かせてたんだろうかね。

金子　わかります。

小林　あと軍モノで面白いのでいえば、オーストラリアかどこかの発着誘導員用のベスト(※8)。甲板で目立つように色が派手で。コットンだから古い時代のものだと思う。

金子　うわぁ、ずっと見てしまう。古着屋さん巡っても全然見掛けませんもん。

小林　いわゆるキャンプベスト(※9)も面白いんだよ。ハイキング用でメーカーもいろいろなんだけど、パイピングやポケットの配置が

Kobayashi: Taking inspirations from them, we designed a vest together. Every pocket has stamped words, but we chose the words with a little twist. For example, words like "songs", "sound" and "voices" are stamped on a pocket for a music player.

Kaneko: That's a fantasy as well. Have you ever put so many words on your product when designing clothes?

Kobayashi: No, no. But I'm glad it worked out.

Kaneko: The vest still looks so cool. I think it was ahead of its time. You also have a lot of military vests, don't you?

Kobayashi: A vest which was made around when the Normandy Landings took place.

Kaneko: This one seems to be appeared in the movie "Saving Private Ryan".

Kobayashi: Yes, it's a typical so-called assault vest. Also, as for the one called a tool vest, I like this British Army's one (*7) which can store banana magazines for a machine gun. It was made between the 1930s and 1940s but still has basting threads.

Kaneko: That's amazing. The curves are also unique.

Kobayashi: Military clothes made by the British Army around World War II are somehow like handicrafts. It's totally opposite of American jeans. I wonder if their tailoring culture was still prevalent back then.

Kaneko: I know what you mean.

Kobayashi: Speaking of unique military vests, this marshaller vest (*8) made in Australia or somewhere else is interesting. The color combination is so bold to be easily found on the deck. I think it's quite old because it's made of cotton fabric.

Kaneko: Wow, I can't take my eyes off. You can never find things like this at a vintage clothing shop.

Kobayashi: Camp vests (*9) are interesting as well. Though they were made for hiking by different manufacturers, the piping and pockets are all so similar.

全部似ていて。

金子 これハイキング用だったんだ。めちゃめちゃ似てますね。不思議。僕のお気に入りは、まずエディ・バウアーのデコイバッグベスト(※10)ですね。カモのおとりを入れるためのポケットがいっぱい付いている。

小林 衝撃的だよなぁ。おとりを入れておくの？ で、おとりを水に浮かべて自分は茂みに隠れると。だからダックハンター柄なのか。

金子 はい。デコイは12個入ります。

小林 狩りにはデコイのほかに散弾銃を持って、弾を入れるバッグを持って、双眼鏡も持って……となると、この着るカタチが最善ってことなのかなぁ。全部想像だけど(笑)。

金子 収納系だと、パタゴニアのパックベスト(※11)も好きで。これ、じつはバッグを分離しても使える3wayなんです。

小林 イイねこれ面白い！ 何年のもの？

金子 1996年ですね。

小林 初めて見たけどよく出来てる。単年の商品だったんだろうか。これをモチーフに何か作ったら怒られちゃうかな(笑)。

金子 小林さんと作ってみたいですけどね。あと今日持ってきたのはマスランドのベスト(※12)。これも大きく3つのパーツからなっているんですが、もしコレを作れっていわれたらややこしそうですね(笑)。

小林 大変だろうなぁ。でも作ってみたいな。もうデザインのやり口から違うもん。

金子 いいデザイナーがいたんですかね？

小林 いい意味で軽く変態な人がいたんだろうね。才能だと思うもん。1つのブランドのアイテムが、これだけ他と違うんだからさ。にしてもこのベスト、バータック(閂)がまた凄いなぁ。どれだけモリッ！としてるのよ。

金子 いや、そこを見ますか(笑)。やっぱり作り手の目線は違うな。さすがです。

小林 いやいや、そんなことはないよ。ちなみに今日見せてくれたベストは、要らなくなったらいつでも俺が引き取るからね(笑)！

Kaneko: So they are made for hiking. That's strange that the designs are so similar. One of my favorites is Eddie Bauer's decoy bag vest (*10). There are many pockets on it to store duck decoys.
Kobayashi: That's surprising. Do you put decoys in it? You then float a decoy down the stream and hide in the bushes. That's why the vest comes in a duck hunter pattern, right?
Kaneko: Yes, you can carry 12 decoys at once.
Kobayashi: Other than decoys, you have to carry a shotgun, bullet bag, binoculars and more for hunting. So is wearing the best way to carry decoys? It's all just my imagination, though.
Kaneko: Patagonia's bag vest (*11) is another wearable storage I like. Because the bag is detachable, you can use it in 3 different ways.
Kobayashi: Nice and interesting! When was it made?
Kaneko: It was made in 1996.
Kobayashi: I haven't seen this before. It's really well made. I wonder if they offered this just for one year. Do you think I'll be in trouble if I design something based on this one?
Kaneko: I'd love to do it with you. And I also brought a vest from Masland (*12) today. It roughly consists of 3 parts, but the construction is so complicated. If we were asked to make this, it would be a hassle.
Kobayashi: It must be. But I'd like to give it a try. Their designing approach is completely different from ours.
Kaneko: I wonder if they had a talented designer.
Kobayashi: I guess there was a designer who was a bit eccentric in a good way. It's a kind of a talent, I think. It's amazing a product made by the brand is so distinct from the ones made by others. By the way, bartack stitches on this vest are so huge and impressive.
Kaneko: Did you detect that? You definitely have designer's eyes. That's as great as I expected.
Kobayashi: No, no. But, you know, I can take all the vests you brought today any time if you no longer need them!

● THE OTHER MY COLLECTIONS

L.L.Bean × BEAMS PLUS 83 DEEP BOTTOM BOAT AND TOTE

1944年、氷や木材を運ぶ鞄として誕生したBean's Ice Carrierを原点に持つBoat and Tote。ビームス プラスの先輩達が足繁く交渉を重ね、2014年に初めて型から別注したモデルであるDeep Bottomが誕生しました。今でもメイン州で作られているそれは、スペックとデザインを現代的にアップデートした温故知新な銘品。以来、配色を変えながら作り続けている思いの深い品です。

Boat and Tote was originally developed in 1944 as Bean's Ice Carrier to carry ice and wood. After repeated negotiations with L.L.Bean, my predecessors at BEAMS PLUS succeeded to launch our completely exclusive model, Deep Bottom, in 2014. Since then, it has still been made in Maine by continuously updating the company's masterpiece with a modern twist and changing the color combination. It's definitely a memorable piece for me and BEAMS PLUS.

L.L.Bean × BEAMS PLUS 84 BEAN BOOTS

通称ビーン・ブーツ。正式名をメイン・ハンティング・シューズ。創業者が、1912年に発明したこの沼地用シューズを、ビームス プラスで2013年に初めて別注し、以来アイデンティティを大切にしながら定期的に別注を続けてきました。エルエルビーンのコーポレートカラーやスエードで別注したり、バックルが付いたラウンジ履きのアーカイブモデルを復刻したり。雨の日にこのシューズを履けばアメリカンムード満点ですね。

These boots are officially named as Maine Hunting Shoes but known better as Bean Boots. In 2013, BEAMS PLUS collaborated with L.L.Bean to first launch the exclusive model of the legendary shoes invented by the founder of the company to walk in a muddy field. And the collab models have regularly been produced since then, from the one in L.L.Bean's corporate colors to the one made of suede to the slipper version with buckles. These boots add an extra American vibe to your rainy-day outfit.

● THE OTHER MY COLLECTIONS

GREGORY × KAPTAIN SUNSHINE × BEAMS PLUS

85 CASSIN

誰が言ったか、バックパック界のロールスロイス。グレゴリー40周年を記念にトリプルネームで作った、シリアルナンバー入りの100個限定作です。1977年のモデルをベースに、カラーをネイビーにしたり初期ロゴを復活させたり。作りが複雑すぎて、精鋭が集うサンプルレーンで生産したのは裏話。115Lと大容量のため、出張にも活躍。

Some say Gregory's Cassin is the Rolls Royce of backpacks. To celebrate the 40th anniversary of the brand, BEAMS PLUS created this special model in a collaboration with them and Kaptain Sunshine for limited 100 pieces only. Designed based on the backpack made in 1977, it comes in navy color with the original logo. The construction is so complicated that the prototype had to be made on a production line with skilled workers. With 115L volume, it is ideal for a business trip.

86 SPORTS SHIRTS

頑丈なことからロッククライミングなどのアメリカンアウトドアシーンで着用された、ラガーシャツの兄弟的存在。とにかく配色が最高で、コレクションするほどに色の魅力に取り憑かれます。50〜60年代のカタログを見ると、ホッケー用やフットボール用などがあり、種目ごとにパターンも様々。ちなみに写真の黒いシャツは、現代と同様に審判用。

Because of the durability, these sports shirts used to be worn for various outdoor activities such as rock-climbing in the U.S. Each of them looks like a rugby shirt and is characterized by the excellent color combination. The more I collect the shirts, the more I become fascinated by the colors. The catalogs from the 1950s to 1960s say there were shirts made for hockey and football in specialized design. The black ones in the photo are made for referees.

● THE OTHER MY COLLECTIONS

87 PATCHES

アウトドアアイテムにもよく見られるパッチの源流は、ヨーロッパの貴族の紋章。おそらくこれが転じて、チームなどの所属を表現するためのものとなり、企業のコマーシャルにも用いられるようになったのではないでしょうか。中央はパッチ文化を象徴するかのような、ボーイスカウトベストです。70年代はパッチが充実していた時代で、個性派揃い。グラフィックにアメリカらしさがあふれています。

Patches are often attached on outdoor clothes in the modern era, but it was originally created for coats of arms in Europe. I guess they later changed their role and became the symbols of teams, organizations and corporations. A garment in the center is a Boy Scout vest that represents the "patch culture". A variety of patches were created especially in the 1970s and each of the graphics conveys the mood of America in a unique way.

126

Recreational Equipment, Inc. 88 PONCHO

REI製、80年代のポンチョです。シンプルな作りなんだけど、いや、シンプルなだけに着たときのドレープ感が素晴らしくて。そして何よりツール感が男心をくすぐります。僕はちょっとした雨なのに傘を差すのは面倒なたち。雨の日にこれを被ると、アメリカっぽいなぁと気分に浸る自分が…。バックパックを背負ったまま被れるのも便利です。

This poncho was made in the 1980s by REI. Because it is simply made, the drapes look beautiful when you wear it. And I love the gadget-like feel. I'm a kind of person who cannot be bothered to use an umbrella for just light rain, and wearing this on a rainy day makes me feel as if I'm in the U.S. You can wear this over a backpack.

● THE OTHER MY COLLECTIONS

SIERRA DESIGNS × BEAMS PLUS 89 EXCLUSIVE PRODUCTS

シエラデザインズには、毎シーズンというほど別注を仕掛けています。60/40クロスのマウンテンパーカしかりミナレットパーカしかり、カグールしかり。名作をベースに生地をより軽いものに変えたり防水性を高めたりと、機能的な観点はもちろんのこと、ファッション観点においてもその時代に合わせたアップデートを加えてきました。お気に入りが増えるばかりです。

We collaborate with Sierra Designs almost every season to make exclusive products. What we have made together includes 60/40 cloth mountain parkas, minaret parkas and cagoules. We have updated the brand's masterpieces not just from a functional perspective such as lightness or waterproof properties, but also from a fashion perspective in line with the times. My list of favorite garments has thus been increasing.

90 DOWN SHOES
91 DOWN HOOD

ダウンシューズは、テントや寝袋内などで履くラウンジスリッパ。ふっくらしたフォルムや配色の妙に惹かれて、いいのを見つけては買ってしまいます。家でも冬場のデスク仕事のときなどに履いています。昔から頭寒足熱っていいますからね。ダウンフードは寝袋とドッキングして保温性を高めるニッチなギア。袖を通すと、アメフトの選手みたいなシルエットになります(笑)。

Down shoes are designed to be worn like slippers in a tent or sleeping bag. The plumpy shape and unique color combinations always fascinate me. Whenever I find a nice pair, I cannot help myself getting it. I wear it when doing desk work at home in wintertime, as it is said that it's good to keep your feet warm. This down hood is made to be attached to a sleeping bag for better thermal performance. You may look like an American footballer in it .

● THE OTHER MY COLLECTIONS

Eddie Bauer 92 DOWN FACE MASKS

ダウンの雄、エディーバウアーならではのニッチなアイテム。アラスカなどの極寒地で役立つ防寒用のマスクです。背面がニットになっており、被ると保温力が高まる秀逸な作りに脱帽。ブランドのカタログにも長い間登場していましたが、希少なアイテムなのでここ最近注目されています。マスクって子どもの頃から男の子の憧れですよね。

Down jacket giant Eddie Bauer once offered these down face masks to ward off the frost in Alaska and other extremely cold areas. The design is well-considered with the knitted lining that keeps the face even warmer when being worn. The item had long been featured in the brand's catalogs, and is now in the spotlight again because of the rarity. Masks are every little boys dream, aren't they?

93 BALACLAVAS

バラクラバ、目出し帽です。クリミア戦争時の兵士の防寒具がルーツともいわれますが、第二次世界大戦時でもミリタリーギアとして用いられていました。スキー用、登山用とさまざまなタイプがあり、キャラが立ったのを見つけるとつい買ってしまう。左下のエルエルビーンのモデルは、柔らかい鹿革かつストラップ構造で着脱しやすい仕様。ギア感がたまりません。

The name, balaclava, said to derive from knitted headgear worn by soldiers in the Crimean War, and the item was also used by military troops during the World War II. There are various types of balaclavas made for a specific purpose such as skiing and climbing. Whenever I find unique and weird ones, I end up buying them. The one on the bottom left was designed by L.L.Bean and is made of soft deer skin. It's also easy to put on and take off with straps. I love the gear-like design.

● THE OTHER MY COLLECTIONS

94 MOCCASIN SHOES

モカシンのルーツはアメリカの先住民が履いていた、スリッポンタイプの一枚革の靴。アウトドアスタイルに欠かせない靴なので何十足も持っています。ラッセルモカシン、アローモカシンなどさまざまなブランドのモカシンがありますが、ランコートはビームス プラスでも長年取り組みをしています。工場を訪れた際に、目の前で見た職人のモカ縫いは、息を呑むものがありました。

The origin of moccasin shoes is Native American's slip-on shoes made of a single piece of leather. Because it is essential for outdoorsy outfits, I own dozens of pairs. Though there are so many brands offering moccasin shoes from Russell Moccasin to Arrow Moccasin, BEAMS PLUS has a quite long relationship with Rancourt & Co.. When I saw their craftsmen sewing shoes at their factory, I was so impressed by their skills.

95 CATALOGS

どんなジャンルにもカタログが存在するのが、アメリカ。アウトドアメーカーのカタログを時代ごとに見比べることで、当時の商品構成だけでなく、生地の変遷、ディテールの進化、シーズンカラーの変遷を知ることができます。だからカタログは深堀りに欠かせない教科書であり、宝物。ただ2〜3年は内容が大体同じなので、せっかく買ったのに後悔することも多いですね(笑)。

In the US, every genre has its catalog. By comparing outdoor wear catalogs issued in different eras, you can learn not just the product lineup of the time but also the changes in fabrics, the improvement of detailing as well as the transition of seasonal colors. So, for me, catalogs are precious textbooks to deepen my knowledge. But I often regret after getting new one because most of them feature the same things for a few years .

●THE OTHER MY COLLECTIONS

Frostline Kits 96 KITS

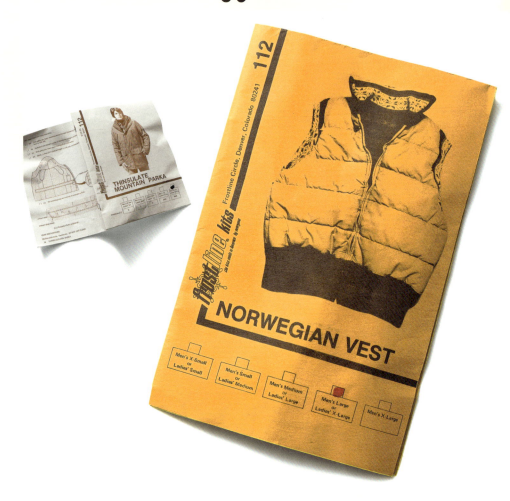

1966年、コロラド州デンバーで創業したフロストラインキッツは、服を完成品でなくキットで販売し成功を収めた草分けブランド。キットにはミシン以外のすべての材料が同封されています。僕もいつか自分で製作したいですね。ちなみに創設者曰く"If you want it done right,do it yourself!(うまく作りたいなら自分でやれ！)"とのこと。

Frostline Kits was established in Colorado in 1966 and known as one of the pioneers of sew-it-yourself kit brands. Their kits included everything other than a sewing machine. I want to make one by myself one day. The founder of the brand said; "If you want it done right, do it yourself!"

● THE OTHER MY COLLECTIONS

Banana Republic 97 TRAVELER'S SPORTS COAT

名作カタログとしても有名な、"トラベル&サファリ"をテーマにしたコレクションの一作です。その名をトラベラーズ スポーツコート。内外に11個のポケットを備え、肘にはエルボーパッチが施された多機能ジャケットの走りといえます。高級エジプト綿を高密度に織ったツイル生地で、しなやか且つエレガントなサファリトリップを。

This jacket was made by Banana Republic for their Travel & Safari collection and featured in their acclaimed catalog. Named as Traveler's Sports Coat, it comes with 11 inner and outer pockets as well as patched elbows. It can be described as the early version of multi-functional jackets. The densely-weaved quality Egyptian cotton fabric feels smooth and takes you to your fabulous trip to safari.

Patagonia 98 BLAZER & BOMBACHAS

ロゴ入りのナイロン生地を裏地に用いたブレザーと、ブーツインできるようプリーツを入れて裾を絞ったチノパン。どちらもパタゴニアらしい素材使いや機能美に心が躍ります。これらで装う当時のスタイルは、さしずめ"マウンテンアイビー"といったところ。

This blazer comes with logoed nylon lining, and these pleated chinos are designed with tapered hems that can easily be tucked in the boots. Both are made by Patagonia and fascinate me with the unique use of fabric and functional beauty. A look with these items can be called as "mountain Ivy" style.

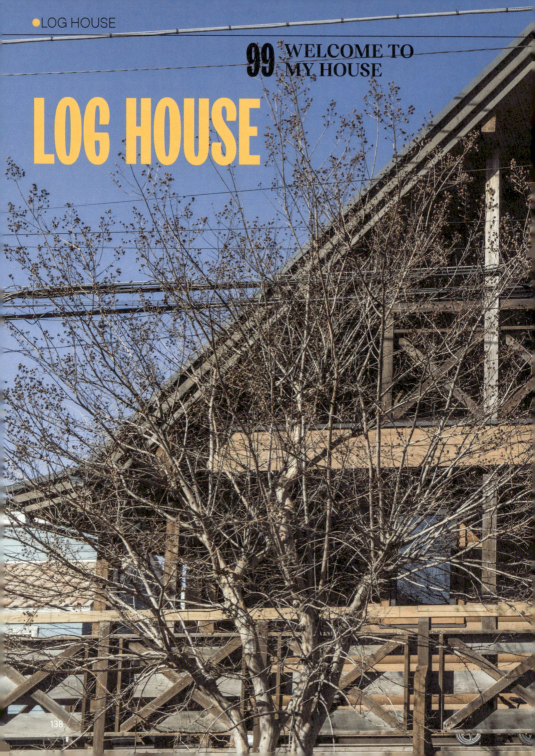

薪ストーブで暖を取り、庭でバーベキューを楽しむ
ログハウスで生涯、EXPEDITION LIFEを!

ログハウスの建築技術は、フィンランドの森の民族、フォレスト・フィン人によってアメリカに伝えられたといわれています。森の生活に憧れ、ログハウスを建てたのは8年前。よく「住み心地はどう?」と聞かれますが、温かいとか寒いとか、これまで住んだ家との違いは正直あまり感じません(笑)。ただひとつ違うのは、家が"生きている"感覚があること。防腐剤を塗るなどの手間は掛かりますが、手入れをしっかり行えば、ログハウスは100年持つともいいます。いわば育てる道具。愛着をもたないはずがありません。子ども達は生まれた時からこの家に住んでいるから、もっと機能的な普通の家に憧れることもあるでしょう。でもこの家の木の香りや温もりは気持ちを落ち着かせてくれるし、壁や床にはたくさんの思い出が残っている。いつかきっと、よさをわかってくれると信じています。

Living EXPEDITION LIFE in a Log House

It is said that the technique of building log houses was brought to America by the Forest Finns, a forest people in Finland. 8 years ago, I have started living in the forest and built a log house. People often ask me, "How is it to live there?", but to be honest, I don't feel much difference from other houses I had lived before. There is no difference in heat or cold, but the only difference is that I feel this house is "alive". It requires some work such as applying preservatives, but if properly maintained, a log house can last for 100 years. So, in a sense, this house is a tool that I can cultivate. How can I not be attached to it? Because my children have lived in this house since they were born, they may yearn for a more functional, usual house. However, the woody scent and warmth of this house calm us down, and the scratches on the walls and floors are filled with memories. I believe that one day they will understand the goodness.

● LOG HOUSE

LOG HOUSE 99 LIVING IN "WOOD"

隅から隅まで木に囲まれた
温かな暮らし

妻もビームス勤務のため、靴が多い我が家。家を建てたとき、シューズラックを自作することは決めていました。自分で塗装を行い、アイアンパーツも好みのクラシックなものを選んでいます。子ども用のデスクは、プレイマウンテンにオーダー。バーベキュー棚のある庭は、ヤードランドスケープの梅津氏に手掛けていただきました。リビングには薪ストーブを設置。冬場はその柔らかな温かさに癒されています。庭での薪割りも楽しいですね。

Warm and cozy life
surrounded by wood

Because my wife also works at BEAMS, there are so many shoes in our house. So I was determined to build a shoe rack by myself when I built our own house. I used iron parts that gave us classic vibe and painted it by myself. As for my children's desks, I ordered them at Playmountain. The garden with barbecue shelves was designed by Mr.Umezu of Yard Landscape. There is a wood stove in our living room and we enjoy the cozy warmth in winter. Chopping firewood in the garden is also fun.

休日はロフトで
くつろぎの時間を

ロフトは、読書などをしてくつろぐ癒しの空間。ここにはニーチェアエックスのロッキングチェアを置いています。座り心地が最高ですね。ダイニングの主役であるテーブルは、ハイクでオーダーしたキャンプデスクをデザインモチーフにしたもので、椅子はミッドセンチュリーの名作たち。ハーマンミラー印のイームズシェルチェアの希少なターコイズカラーは息子が、レモンイエローは娘が選びました。ともにナイスなチョイスだと思います。

Spending days off
Relaxing in the loft

The loft is a relaxing space where we enjoy reading or other activities. I use a rocking chair from NychairX there because it's super comfortable. I ordered our dining table at Hike and the design is based on a camping table. All the chairs around the table are mid-century masterpieces. The colors of Herman Miller's Eames shell chairs were chosen by my children. My son picked rare turquoise, while my daughter went for lemon yellow. Both are excellent choices.

● LOG HOUSE

LOG HOUSE 99 WITH MODERN FURNITURES

家族4人。家のすべてに
"好き"を詰め込んで

壁のアートは、スクラップのスケートボードを用いたジョージ・ピーターソンの作品。5作所有していて、リビングのウッドテーブルは彼の代表作です。その脇はコーア・クリントのサファリチェアで、白のスツールはゲルチョプの"LOG"。キッチンキャビネットは、ハイクでオーダーした妻のお気に入りです。ダウンが山積みの趣味部屋は、将来子ども部屋に。いつかこの部屋からEXPEDITIONしていく姿を温かく見守っていきたいですね。

Family of four living in a house
Full of favorite things

The art piece on the wall was created by George Peterson with scrapped skateboards. I own 5 of his artwork and the wood table in our living room is one of his masterpieces. Kaare Klint's safari chair is beside the table, and there is also a white stool from Gelchop named LOG. My wife's favorite kitchen cabinet was ordered at Hike. My hobby room is now filled with vintage down clothes, but will be given to my children. I want to watch over them warmly until they will leave the room for their own EXPEDITIONS.

142